2nd International Workshop on Theoretical and Computational Physics (IWTCP-2) and 39th National Conference on Theoretical Physics (NCTP-39)

Journal of Physics: Conference Series
Volume 627

Buon Ma Thuot, Vietnam
28 - 31 July 2014

ISBN: 978-1-5108-0799-0
ISSN: 1742-6588

Printed from e-media with permission by:

Curran Associates, Inc.
57 Morehouse Lane
Red Hook, NY 12571

Some format issues inherent in the e-media version may also appear in this print version.

Printed by Curran Associates, Inc. (2015)

For permission requests, please contact the Institute of Physics
at the address below.

Institute of Physics
Dirac House, Temple Back
Bristol BS1 6BE UK

Phone: 44 1 17 929 7481
Fax: 44 1 17 920 0979

techtracking@iop.org

Additional copies of this publication are available from:

Curran Associates, Inc.
57 Morehouse Lane
Red Hook, NY 12571 USA
Phone: 845-758-0400
Fax: 845-758-2634
Email: curran@proceedings.com
Web: www.proceedings.com

Table of contents

Volume 627

IWTCP-2 and NCTP-39
28–31 July 2014, Buon Ma Thuot, Vietnam

Accepted papers received: 3 June 2015
Published online: 26 June 2015

Preface

Papers

2nd International Workshop on Theoretical and Computational Physics (IWTCP-2): Modern Methods and Latest Results in Particle Physics, Nuclear Physics and Astrophysics

39th National Conference on Theoretical Physics (NCTP-39)

Journal of Physics: Conference Series **627** (2015) 011001 doi:10.1088/1742-6596/627/1/011001

Preface

This volume contains selected papers presented at the 2nd International Workshop on Theoretical and Computational Physics (IWTCP-2): Modern Methods and Latest Results in Particle Physics, Nuclear Physics and Astrophysics and the 39th National Conference on Theoretical Physics (NCTP-39). Both the workshop and the conference were held from 28th – 31st July 2014 in Dakruco Hotel, Buon Ma Thuot, Dak Lak, Vietnam.

The NCTP-39 and the IWTCP-2 were organized under the support of the Vietnamese Theoretical Physics Society, with a motivation to foster scientific exchanges between the theoretical and computational physicists in Vietnam and worldwide, as well as to promote high-standard level of research and education activities for young physicists in the country.

The IWTCP-2 was also an External Activity of the Asia Pacific Center for Theoretical Physics (APCTP).

About 100 participants coming from nine countries participated in the workshop and the conference. At the IWTCP-2 workshop, we had 16 invited talks presented by international experts, together with eight oral and ten poster contributions. At the NCTP-39, three invited talks, 15 oral contributions and 39 posters were presented.

We would like to thank all invited speakers, participants and sponsors for making the workshop and the conference successful.

Trinh Xuan Hoang, Nguyen Anh Ky, Nguyen Tri Lan and Nguyen Ai Viet

Journal of Physics: Conference Series **627** (2015) 011001 doi:10.1088/1742-6596/627/1/011001

Organizers

- Institute of Physics, Vietnam Academy of Science and Technology (IOP-VAST)
- University of Science, Vietnam National University, Hanoi (HUS-VNU)

Sponsors

- Asia Pacific Center for Theoretical Physics (APCTP)
- Institute of Physics, Vietnam Academy of Science and Technology (IOP-VAST)
- University of Science, Vietnam National University, Hanoi (HUS-VNU)
- Vietnam Academy of Science and Technology (VAST)

Journal of Physics: Conference Series **627** (2015) 011001 doi:10.1088/1742-6596/627/1/011001

Committees

NCTP-39

Honorary Chair

- Nguyen Van Hieu (Vietnam Academy of Science and Technology, Hanoi)

Chair

- Nguyen Ai Viet (Institute of Physics, VAST, Hanoi)

Organizing Committee

- Nguyen Tri Lan (Institute of Physics, VAST, Hanoi), Chair
- Trinh Xuan Hoang (Institute of Physics, VAST, Hanoi), Co-chair
- Bach Thanh Cong (College of Science, VNU, Hanoi)
- Hoang Dzung (Vietnam National University, Ho Chi Minh City)
- Hoang Anh Tuan (Institute of Physics, VAST)

Program Committee

- Nguyen Toan Thang (Institute of Physics, VAST, Hanoi), Chair
- Phung Van Dong (Institute of Physics, VAST)
- Ho Trung Dung (Ho Chi Minh City Institute of Physics)
- Vu Ngoc Tuoc (Hanoi University of Technology)

IWTCP-2

Chair

- Nguyen Ai Viet (Institute of Physics, VAST, Hanoi)

International Advisory Board

- Guido Altarelli (CERN, Geneva, and Università di Roma Tre, Rome)
- Patrick Aurenche (LAPTH, Annecy)
- Han-Yong Choi (APCTP, Pohang, and Sungkyunkwan University, Suwon)
- Pierre Darriulat (Institute of Nuclear Science and Technology, Hanoi)
- Fadi Ibrahim (Institut de Physique Nucléaire, Orsay)
- Peter Fulde (Max Planck Institute for Physics of Complex Systems, Dresden)
- Pham Quang Hung (University of Virginia, Charlottesville)
- Dao Tien Khoa (Institute of Nuclear Science and Technology, Hanoi)
- Serguey Petcov (SISSA, Trieste)
- Kok-Khoo Phua (NTU and South East Asia Theoretical Physics Association, Singapore)
- Alexei Smirnov (Max-Planck Institute, Heidelberg)

Journal of Physics: Conference Series **627** (2015) 011001 doi:10.1088/1742-6596/627/1/011001

- Masanori Yamauchi (KEK, Tsukuba)

Organizing Committees

- Nguyen Tri Lan (Institute of Physics, VAST, Hanoi), Chair
- Trinh Xuan Hoang (Institute of Physics, VAST, Hanoi), Co-chair
- Bach Thanh Cong (College of Science, VNU, Hanoi)
- Hoang Dzung (Vietnam National University, Ho Chi Minh City)
- Hoang Anh Tuan (Institute of Physics, VAST, Hanoi)

Program Committee

- Nguyen Anh Ky (Institute of Physics, VAST, Hanoi), Chair
- Ha Huy Bang (College of Science, Vietnam National University, Hanoi)
- Phung Van Dong (Institute of Physics, VAST, Hanoi)
- Le Van Hoang (University of Pedagogy, Ho Chi Minh City)
- Nguyen Quynh Lan (Hanoi National University of Education, Hanoi)
- Phan Bao Ngoc (International University, Ho Chi Minh City)
- Dinh Van Trung (Institute of Physics, VAST, Hanoi)

Journal of Physics: Conference Series **627** (2015) 011001 doi:10.1088/1742-6596/627/1/011001

List of Participants

Adam Gheorghe
Joint Institute for Nuclear Research, Dubna
Email: adamg@jinr.ru

Anh Ky Nguyen
Institute of Physics, VAST
Email: anhky@iop.vast.ac.vn

Đinh Thị Hải Hà
Đại học Sư phạm Hà Nội 1
Email: hanhoc164@gmail.com

Bach Thanh Cong
Faculty of Physics, VNU University of Science
Email: congbachthanh@gmail.com

Ban Gilles
Université de Caen Basse-Normandie
Email: ban@lpccaen.in2p3.fr

Bùi Thái Học
Faculty of Physics, VNU University of Science
Email: thaihoc181@gmail.com

Belanger Genevieve
LAPTH, Annecy
Email: belanger@lapp.in2p3.fr

Bui Phuong Thuy
Nam Dinh University of Technology Education
Email: thuy_24784@yahoo.com

Bui Thi Le Quyen
Thai Binh University of Medical and Pharmacy
Email: builequyenytb@gmail.com

Chakraborty Sabyasachi
Indian Association for the Cultivation of Science
Email: tpsc3@iacs.res.in

Chu Thuy Anh
Institute of Physics, VAST
Email: ctanh@iop.vast.ac.vn

Dang Nguyen Dinh
RIKEN, Wakō city
Email: dang@riken.jp

Journal of Physics: Conference Series **627** (2015) 011001 doi:10.1088/1742-6596/627/1/011001

Dang Van Quan
Ho Chi Minh City Institute of Physics
Email: vanquan0203@yahoo.com.vn

Dao Thi Thuy Nga
Han Thuyen High School
Email: daothuyngaht@gmail.com

Dao Vong Duc
Institute of Physics, VAST
Email: dvduc@iop.vast.ac.vn

Darriulat Pierre
Institute of Nuclear Science and Technology, Hanoi
Email: darriulat@vinatom.gov.vn

Dinh Nguyen Dinh
Institute of Physics, Hanoi
Email: dndinh@iop.vast.ac.vn

Dinh Van Trung
Institute of Physics, VAST
Email: dvtrung@iop.vast.ac.vn

Do Son
Ho Chi Minh City University of Technology
Email: dns.vn2013@yahoo.com

Do Tuan Long
Faculty of Physics, Hanoi University of Science, Vietnam National University
Email: dotuanlong@hus.edu.vn

Doan Tri Dung
Ho Chi Minh City Institute of Physics
Email: dtdung@hcmip.vast.vn

Duong Dai Phuong
Doan 871, Bo Quoc Phong
Email: vanha318@yahoo.com

Frère Jean-Marie
Université de Libre de Bruxelles, Bruxelles
Email: frere@ulb.ac.be

Gong Jinn-Ouk
Asia Pacific Center for Theoretical Physics (APCTP), Pohang
Email: jinn-ouk.gong@apctp.org

Hoang Anh - Tuan
Institute of Physics, Hanoi
Email: hatuan@iop.vast.ac.vn

Hoang-Do Ngoc-Tram
Ho Chi Minh City University of Pedagogy
Email: ngoctramhd@gmail.com

Hoàng Văn Hưng
Khoa Vật Lý - Trường Đại Học Sư Phạm Thành Phố Hồ Chí Minh
Email: hunghv@hcmup.edu.vn

Hoi Dinh Bui
Department of Physics, National University of Civil Engineering
Email: buidinhhoi@gmail.com

Huỳnh Vĩnh Phúc
Dong Thap University
Email: hvphuc@dthu.edu.vn

Huynh Ngoc Bay
Binh Thanh Dong High School
Email: huynhngocbay@gmail.com

Itoh Yousuke
Research center for the early universe, School of Science, University of Tokyo
Email: yousuke_itoh@resceu.s.u-tokyo.ac.jp

Konno Tomoyuki
Belle-2 collaboration and Tokyo Metropolitan University
Email: konno@hepmail.phys.se.tmu.ac.jp

Kurihara Yoshimasa
KEK, Tsukuba
Email: yoshimasa.kurihara@kek.jp

Le Van Hoang
Ho Chi Minh City University of Pedagogy
Email: hoanglv2005@gmail.com

Le Bin Ho
Ho Chi Minh City Institute of Physics
Email: bin262@gmail.com

Le Hong Khiem
Institute of Physics, VAST, Hanoi
Email: lhkhiem@iop.vast.ac.vn

Le Minh Tan
Department of Physics, Faculty of Natural Science and Technology, Tay Nguyen University
Email: tantaynguyen82@yahoo.com

Le Thi Cam Tu
HCMC University of Pedagogy
Email: letu106@gmail.com

doi:10.1088/1742-6596/627/1/011001

Man Van Ngu
Hung Yen Industrial College
Email: cnttcs2@gmail.com

Mineo Hirobumi
Institute of atomic and molecular science, Taipei
Email: mineo1210@gmail.com

Ngo Gia Vinh
Bac Ninh Department of Education and Training
Email: nguqlkt@yahoo.com

Nguyễn Bá Hưng
Vietnam Military Medical University
Email: hungnb80@yahoo.com.vn

Nguyễn Dương Bộ
Trung tâm vật lý lý thuyết, Viện Vật lý, Hà Nội
Email: nguyenduongbo80@gmail.com

Nguyễn Hồng Sơn
Trường Đại học Công đoàn
Email: atvsld@gmail.com

Nguyễn Thị Thùy Nhung
Institute of Physics, VAST
Email: ntnhung@iop.vast.ac.vn

Nguyễn Thị Thúy Hằng
Đại Học Bách Khoa, ĐHQG HCM
Email: hangthuynguyen2001@yahoo.com

Nguyễn Thị Thu Nguyệt
Đại học Hải Phòng
Email: conguyetdhhp1968@gmail.com

Nguyễn Trí Lân
Institute of Physics, VAST
Email: ntlan@iop.vast.ac.vn

Nguyen Ngoc Ty
Đại học Sư phạm Tp. HCM
Email: nguyenngocty1182@gmail.com

Nguyen Ai Viet
Institute of Physics, VAST
Email: vieta@iop.vast.ac.vn

Nguyen Dinh Nam
Faculty of Physics, Hanoi University of Science, Vietnam National University
Email: dinhnamt2@yahoo.com

Nguyen Nhu Xuan
Học viện Kỹ Thuật Quân Sự
Email: xuannn@mta.edu.vn

Nguyen Quang Bau
College of Science, Vietnam National Univerisity, Hanoi
Email: nguyenquangbau@yahoo.com

Nguyen Quang Hung
Tan Tao University
Email: hung.nguyen@ttu.edu.vn

Nguyen Quynh Lan
Hanoi National University of Education
Email: nquynhlan@hnue.edu.vn

Nguyen Suan Han
ĐHKH Tự Nhiên - ĐHQG Hà Nội
Email: lienbat76@yahoo.com

Nguyen Thi Ha Loan
Hanoi Pedagogical University No.2
Email: hloansp2@gmail.com

Nguyen Thi Kim Thanh
Institute of Physics, VAST
Email: nkthanh@iop.vast.ac.vn

Nguyen Thi Phuong Lan
Hanoi Pedagogical University No.2
Email: lanntphuong@gmail.com

Nguyen Thu Huong
Academy of Air Defense and Air Force, Hanoi
Email: huong146314@gmail.com

Nguyen Thuy Trang
Faculty of Physics, VNU University of Science, Hanoi
Email: trangnguyenphys@gmail.com

Nguyen Trong Hien
Jet Propulsion Laboratory, CALTECH, Pasadena
Email: hien.t.nguyen@jpl.nasa.gov

Nomachi Masaharu
University of Osaka
Email: nomachi@lns.sci.osaka-u.ac.jp

Pham Quang Hung
University of Virginia, Charlottesville
Email: pqh@virginia.edu

Pham Thi Bich Thao
Can Tho University
Email: ptbthao@ctu.edu.vn

Phan Thị Ngọc Loan
Khoa Vật lý, Đại học sư phạm thành phố Hồ Chí Minh
Email: loan2vietnam@gmail.com

Phan Hong Lien
LeQuyDon Academy of Technology
Email: phlien.mta@gmail.com

Phan Van Nham
Institute of Research and Development, Duy Tan University
Email: pvnham@gmail.com

Phan-Bao Ngoc
HCM International University - Vietnam National University
Email: pbngoc@hcmiu.edu.vn

Phạm Tuấn Vinh
Dong Thap University
Email: ptvinh@dthu.edu.vn

Phi Quang Van
Institute of Physics, VAST
Email: pqvan@iop.vast.ac.vn

Phung Duy Khuong
Institute of Physics, VAST
Email: pdkhuong@iop.vast.ac.vn

Sasaki Misao
YITP, Kyoto University, Kyoto
Email: misao@yukawa.kyoto-u.ac.jp

Tanabe Tomohiko
The University of Tokyo
Email: tomohiko@icepp.s.u-tokyo.ac.jp

Tandean Jusak
National Taiwan University
Email: jtandean@yahoo.com

To Thi Thao
Institute of Physics, VAST
Email: ttthao@iop.vast.ac.vn

Tran Nu Hanh
Phan Boi Chau High School
Email: nuhanh88@gmail.com

Tran Cong Phong
CĐGD VN / Trường ĐHSP, ĐH Huế
Email: tcphong@moet.edu.vn

Tran Dinh Tham
Trường Đại học Phạm Văn Đồng
Email: tdtham@pdu.edu.vn

Tran Minh Tien
Institute of Physics, VAST
Email: minhtien@iop.vast.ac.vn

Tran Quoc Ha
Ho Chi Minh City University of Education, Vietnam
Email: tranquocha@gmail.com

Trần Ngọc Hưng
Institute of Physics, VAST
Email: tnhung@iop.vast.ac.vn

Trịnh Xuân Hoàng
Institute of Physics, VAST
Email: hoang@iop.vast.ac.vn

Trinh Lan Hoa
Institute of Physics, VAST
Email: trinhlanhoa@gmail.com

Trương Thị Trân Châu
Khoa Khoa học cơ bản, Đại học Sư phạm Kỹ thuật TP HCM
Email: chauttt@hcmute.edu.vn

Vũ Ngọc Tuấn
Institute of Electronics- Institute of Military Science and Technology.
Email: tuanbackc@yahoo.com.vn

Vi Ho Phong
Hanoi University of Science
Email: hophongmc@gmail.com

Viên Văn Võ
Tay Nguyen University, 567 Le Duan Street, Buon Ma Thuot City, Dak Lak Province, Viet Nam
Email: wvienk16@gmail.com

Vu Ba Dung
Hanoi University of Mining and Geology
Email: vubazung305@gmail.com

Vu Ngoc Tuoc
Hanoi University of Science and Technology
Email: tuoc.vungoc@hust.edu.vn

Vu Toan Thang
Ngo Quyen University
Email: vuthang76@gmail.com

Yuan Tzu Chiang
Institute of Physics, Academia Sinica
Email: tcyuan@phys.sinica.edu.tw

Peer review statement

All papers published in this volume of *Journal of Physics: Conference Series* have been peer reviewed through processes administered by the proceedings Editors. Reviews were conducted by expert referees to the professional and scientific standards expected of a proceedings journal published by IOP Publishing.

Journal of Physics: Conference Series **627** (2015) 012001 doi:10.1088/1742-6596/627/1/012001

Fermion Masses from Six Dimensions and Implications for Majorana Neutrinos

J-M Frère[1], M Libanov[2], S Mollet[1] and S Troitsky[2]

[1] Physique Théorique CP 225, Université Libre de Bruxelles, B-1050 Bruxelles, Belgium
[2] Institute for Nuclear Research of the Russian Academy of Sciences, 60th October Anniversary Prospect 7a, 117312, Moscow, Russia

E-mail: frere@ulb.ac.be

Abstract. In these notes, we review the main results of our approach to fermion masses. The marge mass ratios between fermions, confronted with a unique breaking mechanism leading to vector bosons masses, led us to consider the possibility that they result from the overlap of fermion wave functions. Such overlaps vary indeed very strongly if the observed fermion families in 4 dimensions originate in a single family in 6 dimensions, through localized wave functions. This framework leads in a natural way to large mass ratios and small mixing angles between quarks. What came as a surprise is that if we impose that neutrinos behave as 2-component ("Majorana") particles in 4D, a completely different situation is obtained for them. Instead of diagonal mass matrices, anti-diagonal ones emerge and lead to a generic prediction of combined inverted hierarchy, large mixing angles in the leptonic sector, and a suppression of neutrinoless-double beta decay placing it at the lower limit of the inverted hierarchy branch, a challenging situation for on-going and planned experiments. Our approach predicted the size of the θ_{13} mixing angle before its actual measurement. Possible signals at colliders are only briefly evoked.

1. Introduction

The spectrum of masses and mixing of fermions (with masses extending from 0.5 MeV to 170 GeV for the charged ones, while the light neutrinos have masses below 1 eV) is difficult to account for, and calls, in the basic formulation of the Standard Model, for a wide range of Yukawa couplings to the (Brout-Englert-Higgs) scalar sector. One more striking feature is the weakness of the mixing between families of quarks, opposed to the unexpectedly strong ones in the leptonic sector.

We have studied for some time the possibility that such a spectrum originates in fact from a framework where the 3 known families in 4 (3+1) dimensions originate from a single one in a 6-dimensional world. This work was initally centered on the quark sector, but when extending it to the leptonic sector we realized an unexpected result, namely, insisting on Majorana-like neutrinos in 4D implied naturally large mixings, inverted hierarchies, suppression of the neutrinoless double beta decay to a value towards the minimum allowed in inverted hierarchy, and a measurable (but at the time yet unmeasured) third mixing angle in the leptonic sector θ_{13} .

The following sections constitute an overview of the approach and its progress toward the current description, but for detailed calculations and full Lagrangians of the model, we refer the reader to the original papers.

2. From plane to sphere

The initial formulation of our approach took place with the 2 extra dimensions considered to form a plane [1, 2, 3, 4]. In this plane, fermions, gauge and scalar bosons propagate freely, but a scalar field Φ introduces a topological singularity, as is customary in models with large extra dimensions (see, e.g., Ref. [5] for a review). Indeed, the potential of Φ is chosen such that it develops a vacuum expectation value, and the topology assumes that it behaves like $e^{in\phi}$, where (R, ϕ) are the polar coordinates in the extra-dimensions plane. The field Φ vanishes at the origin (to avoid a multiple determination), and a finite-energy solution is obtained with the help of an auxiliary (axial) field A coupling both to Φ and to the matter fields involved. The fermions coupled to A, Φ can be shown to possess "zero modes" in the 2 extra dimensions, meaning that they lead to massless particles in 4D [6, 7]. The number of such zero modes, for each 6D fermion type, can be shown to be equal to n, the "winding number" of the vortex solution so obtained. There is no special reason, except phenomenology, to set $n = 3$ (although we realized later that a winding equal to 4, for instance, would lead to an unacceptable neutrino phenomenology), but this value allows to generate the 3 observed families of fermions from just one family in 6D.

One major difficulty with the "plane" extra 2D is that the fermions are localized close to the core of the vortex, (and similarly the scalars coupled to the vortex), while no such localization occurs for the gauge bosons. Since the effective coupling in 4D is driven by the overlap of the wave functions, this would result in infinitesimal coupling of fermions to gauge bosons. There are two ways out, one being to invoke some (superconductivity-inspired, for instance [8]) mechanism for localizing the gauge particles, but it proves difficult. The simplest way is then to consider the extra 2 dimensions as spanning a sphere, rather than the infinite plane. Although the implementation of the vortex is slightly more thechnical, the changes in terms of phenomenological results are minimal [9, 10, 11]. A major result was established when investigating the possibility of generating Majorana-type neutrinos in 4 dimensions in the context of our 6 dimensional approach. Not only did we show that this was possible (although there are no Majorana particles in 6D), but it also resulted in a link between the Majorana character and the unusual mixing properties of the leptonic sector [12, 13].

3. Fermion fields and vortex profiles, electroweak symmetry breaking

In 6 dimensions, the Lorentz group is represented on 8-component spinors, which can further (for massless particles) be split into two inequivalent representations, according to the $\pm$ eigenvalues of a matrix Γ_7 (6D chirality). With compactification in mind, the resulting 4-component spinors $\Psi_\pm$ will later appear as 4D Dirac spinors, composed of the more usual L and R chiralities.

$$\Psi = \begin{pmatrix} \psi_{+R} \\ \psi_{+L} \\ \psi_{-L} \\ \psi_{-R} \end{pmatrix} \quad . \tag{1}$$

When coupling the 6D fermions to the winding-3 scalar field, 3 massless modes emerge, labeled $n = 1, 2, 3$. It should be noted that they correspond to *chiral* fermions in 4D. Indeed a single 2-component spinor ψ appears, although it comes in a redundant way (it is present in both the + and − parts of the original spinor). We can in this way generate for each 6D fermion 3 chiral fermions in 4D, corresponding to the 3 families of particles. More precisely, we obtain the left-handed fields:

$$L \sim \sum_n \begin{pmatrix} 0 \\ f_{3-n}(r)\; e^{i(3-n)\phi} \psi_{Ln}(x^\mu) \\ f_{n-1}(r)\; e^{i(1-n)\phi} \psi_{Ln}(x^\mu) \\ 0 \end{pmatrix} \quad , \tag{2}$$

Journal of Physics: Conference Series **627** (2015) 012001 doi:10.1088/1742-6596/627/1/012001

Similarly, we can generate right-handed fermions by adapting the coupling to the vortex field Φ:

$$R \sim \sum_n \begin{pmatrix} f_{n-1}(r)\ e^{i(1-n)\phi} \chi_{Rn}(x^\mu) \\ 0 \\ 0 \\ f_{3-n}(r)\ e^{i(3-n)\phi} \chi_{Rn}(x^\mu) \end{pmatrix}, \tag{3}$$

It must be noted that each of the "zero modes" possesses a very specific dependence in ϕ, the polar angle in the remaining 2D, but also that their radial dependance is very different. This will result, on one side (ϕ dependance) to very specific forms for the mass matrix, and on the other side, to very different masses for the different fermion families.

The electroweak symmetry breaking occurs in a very similar way to the 4-dimensional case, through an explicit Standard Model Scalar (Brout-Englert-Higgs fiels) H. The difference here is that the scalar field is bound to the vortex, and its vacuum expectation value is thus dependant on the extra dimensions. In the present case, we chose a situation where it carries no winding number.

The field assignments are listed in Table 1, note that we included an extra scalar field X, which is used to generate the (small) off-diagonal elements of the mass matrix. For details see [13], note however that instead of introducing products of fields (like $\overline{e_R} H^\dagger X \Phi^* L$) in [13], we could choose to use additional scalar doublets H_i.

Field	Notation	$U(1)_g$	$U(1)_Y$	$SU(2)_w$	$SU(3)_c$
vortex scalar	Φ	$+1$	0	**1**	**1**
BEH boson	H	0	$+1/2$	**2**	**1**
auxiliary scalar	X	$+1$	0	**1**	**1**
quark $SU(2)_W$ doublet	(Q_+, Q_-)	$(3,0)$	$1/6$	**2**	**3**
up-type quark $SU(2)_W$ singlet	(U_+, U_-)	$(0,3)$	$2/3$	**1**	**3**
down-type quark $SU(2)_W$ singlet	(D_+, D_-)	$(0,3)$	$-1/3$	**1**	**3**
lepton $SU(2)_W$ doublet	(L_+, L_-)	$(3,0)$	$-1/2$	**2**	**1**
charged lepton $SU(2)_W$ singlet	(E_+, E_-)	$(0,3)$	-1	**1**	**1**
sterile neutrino singlet	(N_+, N_-)	$(0,0)$	0	**1**	**1**

Table 1. Charge assignments of the fields under the gauge groups of the SM and under $U(1)_g$. For fermions, the two numbers in parentheses correspond to the charges of the components with positive and negative values of the Γ_7 parity, denoted everywhere by "+" and "−" indices, correspondingly.

The respective profiles of the charged fermions and the scalar field H are centered on the vortex, and an approximate picture is given in Figure 1.

4. Fermion masses - the good news: neutrinos are different!

Fermion masses, as announced, result from the convolution of the fermion fields (expressed in 6 dimensions) with the scalar boson field. For charged fermions, the (main) couplings are of the type $e_R H^\dagger L$, and the ϕ integration leads to a diagonal matrix (the scalar field carries no winding number), with strongly hierarchical masses (as it has the strongest overlap with the mass-giving $\langle H \rangle$, the fermion with winding number = 0 gets the largest mass). The winding number n acts as an effective family number f with $f = 3 - n$.

The situation is completely different for neutrinos, if we insist that they behave in 4 dimensions as Majorana (Weyl) fields. The masses of our "light" neutrinos result, through

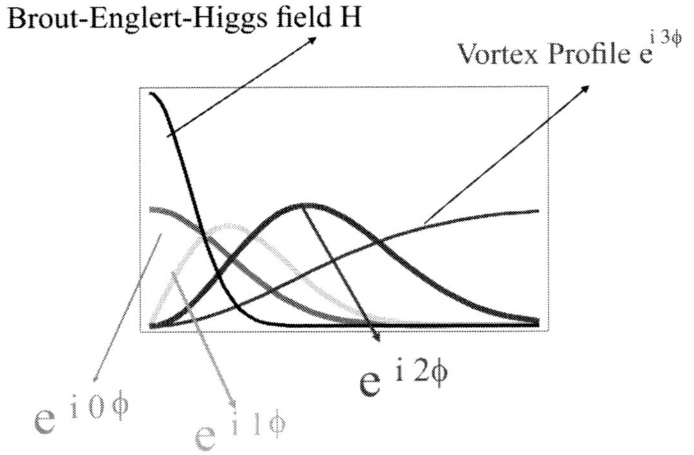

Figure 1. Profiles of the charged fermions, H and vortex field Φ in the extra dimensions.

a see-saw type mechanism, from coupling to the "bulk" field N (the equivalent of the right-handed neutrino in 4 dimensions). A bulk mass term $M\ \overline{N^c}N$ induces a similar contribution for the light neutrino, namely $\overline{\psi^c_\nu}\ \psi_\nu$, where ψ^c is the charge conjugated field. As is readily seen from equation (2) this leads to an off-diagonal mass matrix in family space:

$$\begin{pmatrix} 0 & 0 & m \\ 0 & \mu & 0 \\ m & 0 & 0 \end{pmatrix}, \qquad (4)$$

where the dynamics leads to $\mu \ll m$, while the matrix is symmetrical by construction.

The consequences of this shape are striking. It implies a maximal mixing between neutrinos 1 and 3 (in our notations), but also a degeneracy between the heavier 2 neutrinos.

Of course, this cannot be the whole picture, and extra contributions are needed, both in the charged fermions and neutrino sectors, to provide the full mixing pattern (these smaller contributions are brought in an effective way in our model by the X field); yet, the pattern we have here, after diagonalisation of the neutrino matrix, leading to

$$\begin{pmatrix} m & 0 & 0 \\ 0 & -m & 0 \\ 0 & 0 & \mu \end{pmatrix}, \qquad (5)$$

points to a set of generic predictions:

- inverted hierarchy of neutrinos (2 heavier, closely-spaced neutrinos, and a lighter one);
- large mixing angles (one comes automatically and is linked to the Majorana nature of the 4D neutrinos);
- suppression of the lepton number violation: equal and opposite masses $m, -m$ lead to cancelation of contributions to the neutrinoless-double beta decay: this is a "pseudo-Dirac" scheme, involving neutrinos from 2 families.

This complements the already-made generic features of the quark (and to some extend charged leptons) sectors, namely:

- nearly diagonal mass matrices;
- strong hierarchy of charged fermions masses between the 3 families.

5. Predictions, prospects and experimental constraints

A full description of the masses and mixings necessitates to introduce more (off-diagonal) couplings, and therefore leads to more arbitrariness. We constructed somewhat minimal models (with a limited number of off-diagonal terms) and found reasonable mass spectra and mixings.

One definite and now verified prediction was for a non-zero U_{e3} of order $\sin(\theta_{13}) \simeq 0.13$, before its measurement by the Daya Bay experiment [14].

In terms of checking the Majorana nature of the neutrinos, neutrinoless double beta decay are the only accessible test: the "pseudo-Dirac" partial cancelation puts our prediction at the lower bound of the "inverted hierarchy case", with an effective mass $m_{\beta\beta} \simeq 0.013$ eV which makes it reachable only for a later stage of experiments [15].

In terms of accelerator physics, the prospects are less encouraging, at least in the basic formulation of our approach. Indeed, the extra dimension scale R is associated to the mass of extra gauge bosons in 4 dimensions (Kaluza-Klein excitations). In our case, this involves Z'_0, $Z'_{\pm 1}$ (and similar states for the photon and gluons). The $Z'_{\pm 1}$ cary on unit of winding, which means that they are flavour changing, but conserve (in the first, diagonal matrix, approximation) family number.

A clean test is given by $K^0_L \to \bar{e}\mu$, which provides a limit

$$\kappa/M^2_{Z'} < (100\mathrm{TeV})^{-2}. \tag{6}$$

Here, κ measures the overlap between the wave functions of the fermions pertaining to different families. In our present analysis (we are working on alternate cases), its value does not allow to bring the $1/R$ scale much below 100 TeV, but it is conceivable that other geometries would allow it. In that case $\bar{\mu}e$ (in excess by one order of magnitude over $\mu\bar{e}$) would provide a particularly striking signal [10, 11].

Acknowledgments

This work is funded in part by IISN and by Belgian Science Policy (IAP "Fundamental Interactions"). The work of ML and ST (elaboration of the model of the origin and hierarchy of neutrino masses and mixings in the context of new experimental data) is supported by the Russian Science Foundation, grant 14-22-00161.

References

[1] M. Libanov and S. V. Troitsky, *Three fermionic generations on a topological defect in extra dimensions*, *Nucl.Phys.* **B599** (2001) 319–333, hep-ph/0011095.

[2] J. Frere, M. Libanov, and S. V. Troitsky, *Three generations on a local vortex in extra dimensions*, *Phys.Lett.* **B512** (2001) 169–173, hep-ph/0012306.

[3] J. Frere, M. Libanov, and S. V. Troitsky, *Neutrino masses with a single generation in the bulk*, *JHEP* **0111** (2001) 025, hep-ph/0110045.

[4] M. V. Libanov and E. Y. Nougaev, *Towards the realistic fermion masses with a single family in extra dimensions*, *JHEP* **0204** (2002) 055, hep-ph/0201162.

[5] V. A. Rubakov, *Large and infinite extra dimensions: An Introduction*, *Phys. Usp.* **44** (2001) 871 [*Usp. Fiz. Nauk* **171** (2001) 913]

[6] R. Jackiw and P. Rossi, *Zero Modes of the Vortex - Fermion System*, *Nucl.Phys.* **B190** (1981) 681.

[7] E. Witten, *Superconducting Strings*, *Nucl.Phys.* **B249** (1985) 557–592.

[8] G. R. Dvali and M. A. Shifman, *Domain walls in strongly coupled theories*, *Phys. Lett.* B **396** (1997) 64 [Erratum-ibid. B **407** (1997) 452]

[9] J. Frere, M. Libanov, E. Nugaev, and S. V. Troitsky, *Fermions in the vortex background on a sphere*, *JHEP* **0306** (2003) 009, hep-ph/0304117.

[10] J. Frere, M. Libanov, E. Nugaev, and S. V. Troitsky, *Flavor violation with a single generation*, *JHEP* **0403** (2004) 001, hep-ph/0309014.

[11] J. Frere, M. Libanov, E. Y. Nugaev, and S. V. Troitsky, *Searching for family number conserving neutral gauge bosons from extra dimensions*, *JETP Lett.* **79** (2004) 598–601.

[12] J.-M. Frere, M. Libanov, and F.-S. Ling, *See-saw neutrino masses and large mixing angles in the vortex background on a sphere*, *JHEP* **1009** (2010) 081, arXiv:1006.5196.

[13] J. M. Frère, M. Libanov, S. Mollet and S. Troitsky, *Neutrino hierarchy and fermion spectrum from a single family in six dimensions: realistic predictions* JHEP **1308** (2013) 078, arXiv:1305.4320 .

[14] F. P. An *et al.* [DAYA-BAY Collaboration], "Observation of electron-antineutrino disappearance at Daya Bay," Phys. Rev. Lett. **108** (2012) 171803 arXiv:1203.1669

[15] Anatoly A. Smolnikov, for the GERDA Collaboration, Status of the GERDA experiment aimed to search for neutrinoless double beta decay of 76-Ge arXiv:0812.4194

Dark $U(1)$

Chia-Feng Chang[1], **Ernest Ma**[2] **and Tzu-Chiang Yuan**[3]
[1] Department of Physics, National Taiwan Normal University, Taipei 116, Taiwan
[2] Department of Physics and Astronomy, University of California, Riverside, California 92521, U.S.A.
[3] Institute of Physics, Academia Sinica, 128 Academia Road, Section 2, Nangang, Taipei 11529, Taiwan

E-mail: `60141014S@ntnu.edu.tw`, `ma@phyun8.ucr.edu`, `tcyuan@phys.sinica.edu.tw`

Abstract. In this talk we will explore the possibility of adding a local $U(1)$ dark sector to the standard model with the Higgs boson as a portal connecting the visible standard model sector and the dark one. We will discuss existing experimental constraint on the model parameters from the invisible width of Higgs decay. Implications of such a dark $U(1)$ sector on phenomenology at the Large Hardon Collider will be addressed. In particular, detailed results for the non-standard signals of multi-lepton-jets that arise from this simple dark sector will be presented.

1. Introduction

Although we all cheered triumphantly the standard model (SM) with the recent discovery of the 126 GeV Higgs-like boson [1, 2], precision measurements of the Higgs properties are necessary in order to differentiate whether this is indeed the SM Higgs or a mixture with other scalar bosons in an extended Higgs sector. Such measurements may not be feasible at the Large Hadron Collider (LHC) but can nevertheless be achieved by the future electron-positron machine like the proposed circular electron-positron collider (CEPC) in China or the International Linear Collider (ILC) in Japan. We will present one simple $U(1)$ extension [3] of the SM to demonstrate that the 126 GeV boson could be a mixture and detailed how studies of its non-standard decay modes at the LHC may lead to the discovery of a hidden dark sector with light particle content.

2. Local Dark $U(1)$

In [3], we extend the electroweak SM by including the original Abelian Higgs model for a dark $U(1)_D$. Similar works can be found in [4, 5, 6, 7, 8, 9]. The bosonic part of the Lagrangian density is

$$\mathcal{L}_{\rm B} = \mathcal{L}_{\rm gauge} + \mathcal{L}_{\rm scalar} \tag{1}$$

with

$$\mathcal{L}_{\rm gauge} = -\frac{1}{4}\vec{W}_{\mu\nu}\cdot\vec{W}^{\mu\nu} - \frac{1}{4}B_{\mu\nu}B^{\mu\nu} - \frac{1}{4}C_{\mu\nu}C^{\mu\nu} - \frac{\epsilon}{2}B_{\mu\nu}C^{\mu\nu} , \tag{2}$$

$$\mathcal{L}_{\rm scalar} = |D_\mu\Phi|^2 + |D_\mu\chi|^2 - V_{\rm scalar}(\Phi,\chi) , \tag{3}$$

and

$$D_\mu\Phi = \left(\partial_\mu + ig\frac{1}{2}\sigma_a W_{a\mu} + i\frac{1}{2}g' B_\mu\right)\Phi , \tag{4}$$

$$D_\mu\chi = (\partial_\mu + ig_D C_\mu)\,\chi , \tag{5}$$

where $\vec{W}^\mu$, B^μ and C^μ are the gauge potentials of the $SU(2)_L$, $U(1)_Y$ and $U(1)_D$ with gauge couplings g, g' and g_D respectively, and ϵ is the kinetic mixing parameter between the two $U(1)$s [10]. The scalar potential in (3) is given by

$$V_{\rm scalar} = -\mu_\Phi^2 \Phi^\dagger \Phi + \lambda_\Phi \left(\Phi^\dagger\Phi\right)^2 - \mu_\chi^2 \chi^* \chi + \lambda_\chi (\chi^*\chi)^2 + \lambda_{\Phi\chi} \left(\Phi^\dagger\Phi\right)(\chi^*\chi) \; . \tag{6}$$

We pick the unitary gauge and expand the scalar fields around the vacuum

$$\Phi(x) = \frac{1}{\sqrt{2}} \begin{pmatrix} 0 \\ v + h(x) \end{pmatrix} \quad , \quad \chi(x) = \frac{1}{\sqrt{2}} \left(v_D + h_D(x)\right) \tag{7}$$

with the vacuum expectation values v and v_D fixed by minimization of the potential to be

$$v^2 = \frac{\mu_\Phi^2 - \frac{1}{2}\frac{\lambda_{\Phi\chi}}{\lambda_\chi}\mu_\chi^2}{\lambda_\Phi - \frac{1}{4}\frac{\lambda_{\Phi\chi}^2}{\lambda_\chi}} \quad , \quad v_D^2 = \frac{\mu_\chi^2 - \frac{1}{2}\frac{\lambda_{\Phi\chi}}{\lambda_\Phi}\mu_\Phi^2}{\lambda_\chi - \frac{1}{4}\frac{\lambda_{\Phi\chi}^2}{\lambda_\Phi}} \; . \tag{8}$$

The mass matrix M_S^2 for the scalar bosons is

$$M_S^2 = \begin{pmatrix} 2\lambda_\Phi v^2 & \lambda_{\Phi\chi} v v_D \\ \lambda_{\Phi\chi} v v_D & 2\lambda_\chi v_D^2 \end{pmatrix} . \tag{9}$$

Its eigenvalues are

$$m_{1,2}^2 = \frac{1}{2}\left[{\rm Tr} M_S^2 \pm \sqrt{({\rm Tr} M_S^2)^2 - 4\,{\rm Det} M_S^2} \right] . \tag{10}$$

The physical Higgs (h_1, h_2) are related to the original (h, h_D) as

$$\begin{pmatrix} h_1 \\ h_2 \end{pmatrix} = \begin{pmatrix} \cos\alpha & \sin\alpha \\ -\sin\alpha & \cos\alpha \end{pmatrix} \begin{pmatrix} h \\ h_D \end{pmatrix} , \tag{11}$$

with the mixing angle

$$\sin 2\alpha = \frac{2 m_{12}^2}{m_1^2 - m_2^2} \quad , \quad m_{12}^2 = \lambda_{\Phi\chi} v v_D \; . \tag{12}$$

We will identify the heavier Higgs h_1 with mass $m_1 = 126$ GeV as the new boson observed at the LHC [1, 2], while the lighter one h_2 has been escaped detection thus far. The SM Higgs couplings with the SM fermions and gauge bosons are thus modified by a factor of $\cos\alpha$.

3. Constraints

We had computed the following non-standard processes $h_1 \to \gamma_D\gamma_D$, $h_1 \to h_2 h_2$, $h_1 \to h_2 h_2^* \to h_2 \gamma_D \gamma_D$ and $h_1 \to h_2 h_2 h_2$. Each of the h_2 in the final state of these processes will decay into two dark photons and each dark photon will give rise to two leptons through its mixing with the photon We note that h_2 can decay to SM fermion or gauge boson pairs as well through its mixing with h_1 and hence these modes are suppressed. We take the branching ratio of $h_2 \to \gamma_D\gamma_D$ to be 100%. These non-standard processes will provide multiple leptons in the final state of the standard model Higgs decay [7]. The contribution to the heavier Higgs width from these non-standard processes is

$$\Gamma_{h_1}^{NS} = \sin^2\alpha \hat{\Gamma}(h_1 \to \gamma_D\gamma_D) + \Gamma(h_1 \to h_2 h_2) + \Gamma(h_1 \to h_2\gamma_D\gamma_D) + \Gamma(h_1 \to h_2 h_2 h_2) + \cdots \tag{13}$$

The four lepton modes from the first term $h_1 \to \gamma_D\gamma_D$ followed by $\gamma_D \to l\bar{l}\,(l = e, \mu)$ were studied in details in [4]. Thus the total width of the heavier Higgs h_1 is modified as

$$\Gamma_{h_1} = \cos^2\alpha \hat{\Gamma}_h + \Gamma_{h_1}^{NS} \; , \tag{14}$$

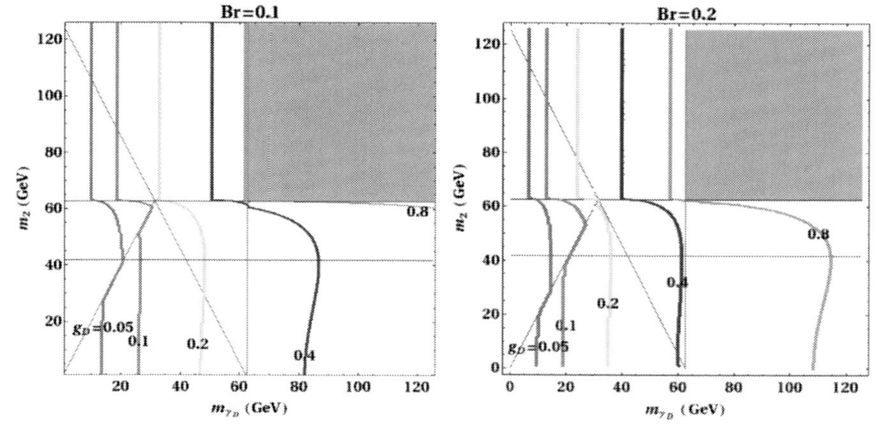

Figure 1. Contour plot of the non-standard branching ratio $B_{h_1}^{NS} = 0.1$ (left) and 0.2 (right) of the heavier Higgs h_1 in the (m_{γ_D}, m_2) plane up to 126 GeV in both axes for $\sin^2\alpha = 0.0009$ and $g_D = 0.05$, 0.1, 0.2, 0.4 and 0.8.

where $\hat{\Gamma}_h$ is the width of the SM Higgs h, which has a theoretical value of 4.03 MeV. The branching ratio for the non-standard modes of the heavier Higgs decay is

$$B_{h_1}^{NS} = \frac{\Gamma_{h_1}^{NS}}{\Gamma_{h_1}} , \tag{15}$$

which should be constrained to be less than 22% or so from global fit results [11].

Now the dark Higgs h_2 decays into $\gamma_D\gamma_D$ and SM particles with coefficients $\cos^2\alpha$ and $\sin^2\alpha$ respectively, so its branching fraction into $\gamma_D\gamma_D$ is given by

$$B(h_2 \to \gamma_D\gamma_D) = \frac{\cos^2\alpha\hat{\Gamma}(h_2 \to \gamma_D\gamma_D)}{\cos^2\alpha\hat{\Gamma}(h_2 \to \gamma_D\gamma_D) + \sin^2\alpha\hat{\Gamma}_{h_2}^{\rm SM}} , \tag{16}$$

where $\hat{\Gamma}_{h_2}^{\rm SM}$ is the partial decay width of h_2 into SM particles. Since $\hat{\Gamma}_{h_2}^{\rm SM}$ are suppressed by a factor of $\sin^2\alpha$, the above branching fraction is close to unity. Formulas for the various decay rates can be found in [3].

In our numerical work, we will restrict our interest where both the dark photon and dark Higgs have masses smaller than 126 GeV. In particular, we will pay special attention to the small mass region where their masses are in the range of 0.5 to a few GeV. In this range, final states of τ pair and light quarks pairs (pion and kaon pairs) from the dark photon decay are also possible, but they are harder to detect at the LHC.

In Fig. 1, we plot the contour of the non-standard branching ratio $B_{h_1}^{NS} = 0.1$ (left) and 0.2 (right) of the heavier Higgs h_1 in the (m_{γ_D}, m_2) plane up to 126 GeV in both directions with the following parameter input: $\sin^2\alpha = 0.0009$ and $g_D = 0.05$, 0.1, 0.2, 0.4 and 0.8.

In Fig. 2, we plot the contour of the non-standard branching ratio $B_{h_1}^{NS} = 0.1$ (left) and 0.2 (right) of the heavier Higgs h_1 in the (m_{γ_D}, m_2) plane for the small mass region of 0.5 to 5 GeV in both axes for $\sin^2\alpha = 0.0009$ and $g_D = 0.005$, 0.009, 0.013 and 0.017.

4. Signals of Multi-Lepton-Jets

Next, we will present our study of some collider signatures for the model. In particular, we will focus on the 4 lepton-jet and 2 lepton-jet modes in our analysis. We consider the following four processes which may lead to signals of multi-lepton-jets at the LHC:

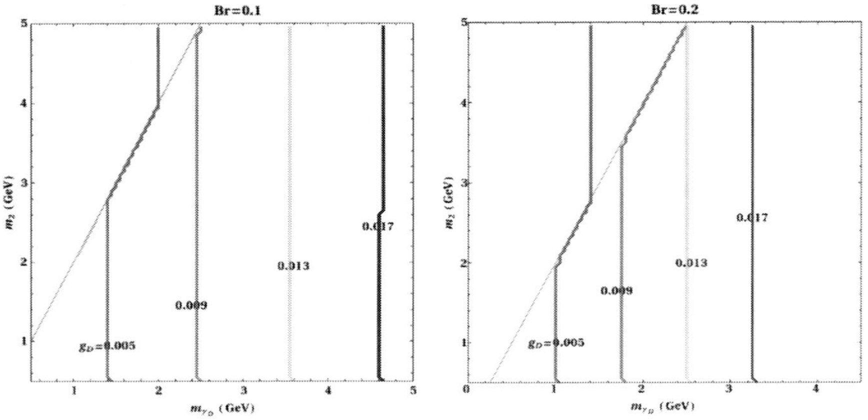

Figure 2. Contour plot of the non-standard branching ratio $B_{h_1}^{NS} = 0.1$ (left) and 0.2 (right) of the heavier Higgs h_1 in the small mass region of 0.5 to 5 GeV in the (m_{γ_D}, m_2) plane for $\sin^2\alpha = 0.0009$ and $g_D = 0.005$, 0.009, 0.013 and 0.017.

$$
\begin{array}{lll}
\text{(I)} & pp \to h \to ZZ \to l^+l^-l^+l^- & \\
\text{(II)} & pp \to VV \to l^+l^-l^+l^- & (VV \; = \; ZZ,\, \gamma\gamma,\, Z\gamma) \\
\text{(III)} & pp \to h_1 \to XX \to l^+l^-l^+l^- & (XX \; = \; ZZ,\, \gamma_D\gamma_D,\, h_2h_2) \\
\text{(IV)} & pp \to h_1 \to h_2h_2 \to \gamma_D\gamma_D\gamma_D\gamma_D \to l^+l^-l^+l^-l^+l^-l^+l^- &
\end{array}
$$

where $l = e$ or μ. Processes (I) and (II) are coming entirely from the SM, process (III) can arise from either SM (with modified Higgs-ZZ coupling) or the dark portal (see Fig. 3), and process (IV) is purely from the dark portal (see Fig. 4).

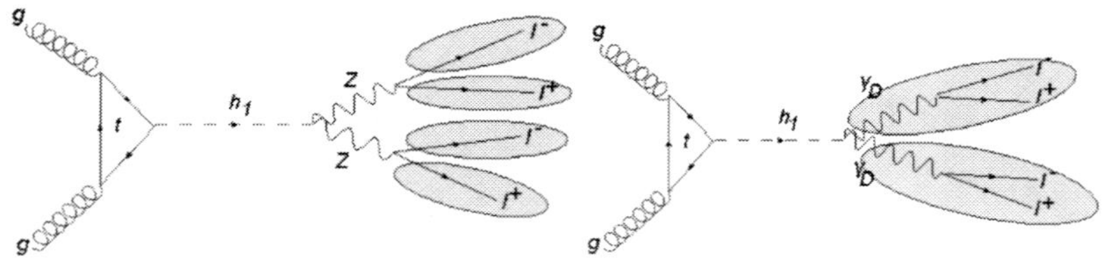

Figure 3. Some topologies of 4 (left) and 2 (right) lepton-jets for process III. The 4 lepton-jets can also be coming from the SM of process I with h_1 replaced by the SM h. The immediate state of h_2h_2 for the 2 lepton-jets is not shown since the branching ratio for $h_2 \to l^+l^-$ is very tiny.

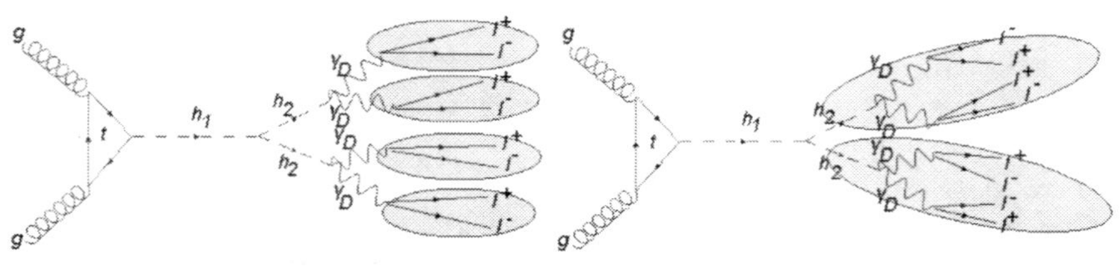

Figure 4. Some topologies of 4 (left) and 2 (right) lepton-jets for process IV.

Journal of Physics: Conference Series **627** (2015) 012002 doi:10.1088/1742-6596/627/1/012002

Table 1. Several benchmark points of the dark portal used to calculate the signals of multi-lepton-jets. ($\epsilon = 10^{-4}$ and $\sin^2\alpha = 10^{-3}$)

Benchmark Point	g_D	M_{γ_D}	m_2	$\mathrm{Br}_{h_1\to\mathrm{DarkStuff}}$	$\mathrm{Br}_{h_2\to\gamma_D\gamma_D}$	$\mathrm{Br}_{\gamma_D\to l^+l^-}$
A	0.005	1.5	4	$\sim 16\%$	99%	50%
B	0.009	1.8	10	$\sim 20\%$	100%	50%
C	0.005	1.5	40	$\sim 15\%$	99%	50%
D	0.005	1.8	40	$\sim 11\%$	99%	50%

In what follows, we will use the 4 benchmark points listed In Table 1 for analysis.

The number of events versus the total invariant mass $M_{\mathrm{Invariant}}$ for the four processes I, II, III and IV at the LHC-14 without any cuts are shown in Fig. 5 for the benchmark point B. We can see that before imposing any cuts the number of events around the Higgs boson resonance for the two processes III (red) and IV (yellow, 8 leptons) from the dark portal can stand above the SM processes of I (blue) and II (black). However away from the resonance region, the 4 leptons SM background from process I (black) is 2 to 3 order of magnitudes above the signals from process IV (green).

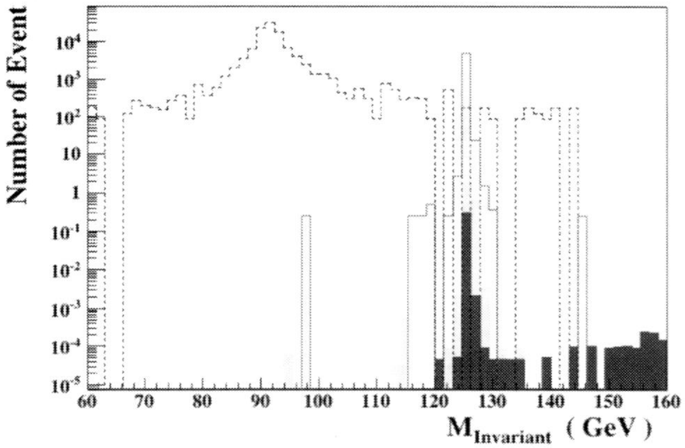

Figure 5. Number of events versus $M_{\mathrm{Invariant}}$ [GeV] without any cuts for benchmark point B at LHC-14 with a fixed luminosity of 10 fb^{-1}. Histogram of blue strip is for process I, black dash is for process II, red solid is for process III, and yellow strip (green dash) is for process IV of 8 (4) leptons.

For the kinematic cuts for the 2 and 4 lepton-jets, we follow Refs. [4, 7] and [12]. For the basic cuts that we will impose in all processes, we have

<u>Basic cuts:</u>

(4 leptons case) $p_{T_l} \geq 20, 10, 10, 10$ GeV, $|\eta_l| < 2.3$;

(8 leptons case) $p_{T_l} \geq 20, 10, 10, 10, 0, 0, 0, 0$ GeV, $|\eta_l| < 2.3$,

where p_{T_l} and η_l are the transverse momenta and pseudo-rapidity of the lepton respectively. On top of the basic cuts, we employ the following lepton-jet cuts

<u>4 lepton-jet cuts:</u> $\Delta R^d_{j_i j_j} > 0.7$, $\Delta R^s_{l_i l_j} < 0.2$, $M_{\mathrm{Invariant}} = M_{h_1} \pm 10$ GeV;

<u>2 lepton-jet cuts:</u> $\Delta R^d_{j_1 j_2} > 0.7$, $\Delta R^s_{l_i l_j} < 0.2$, $M_{\mathrm{Invariant}} = M_{h_1} \pm 10$ GeV.

Here ΔR^d_{jj} denotes the cone radius between two different lepton-jets and ΔR^s_{ll} denotes the cone radius between two different leptons in the same lepton-jet, as depicted in Fig. 6. $M_{\mathrm{Invariant}}$ denotes the invariant mass of all final state particles due to the decay chain of the SM Higgs boson resonance, give or take 10 GeV from the central value of 126 GeV.

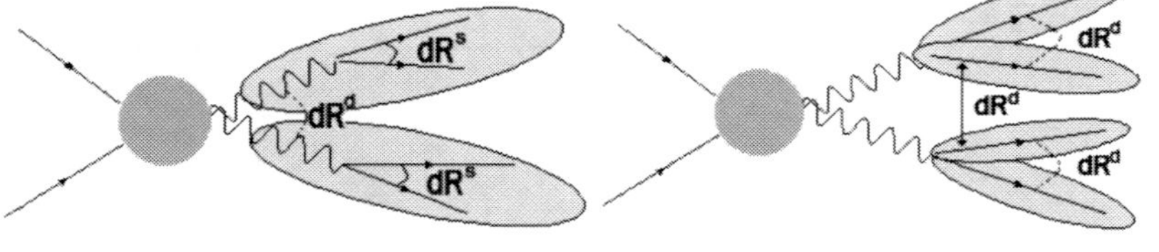

Figure 6. Graphical illustrations for the kinematic cuts on the cone radius ΔR of final state leptons. The 2 and 4 lepton-jet cases are shown in the left and right figures respectively.

We now discuss the impact of imposing the multi-lepton-jet cuts on the cross sections. The topologies of imposing the 4 and 2 lepton-jet cuts for processes III and IV are shown in Figs. 3 and 4 respectively. In Table 2, we show the cross sections of the 4 processes at the LHC-14 with the basic, 4 and 2 lepton-jet cuts for the 4 benchmark points listed in Table 1. The following statements can be drawn from the results shown in Table 2:

- The 4 and 2 lepton-jet cuts have strong and different impact for the SM processes I and II. For process I, since the intermediate state is the Z boson with a relatively high mass, its decay products can be produced at a relatively large angle with respect to the original Z boson direction. Thus it favors 4 lepton-jets in the final state (see left diagram in Fig. 3) and 2 lepton-jets is vanishing small for process I. On the other hand, SM process II has a cross section of about 700 times larger than process I with just the basic cuts imposed. Imposing the 4 and 2 lepton-jet cuts reduce the cross section of process II by a factor of 4.7×10^{-3} and 1.1×10^{-3} respectively. We note that the ZZ intermediate state in process II arises from the tree level parton processes of quark-quark annihilation while in process I it is connected with the loop-induced gluon fusion mechanism of Higgs production.
- For process III since the dark photon mass is small (1.5 GeV for benchmark points A and C, and 1.8 GeV for benchmark points B and D) the contribution from intermediate state of $\gamma_D\gamma_D$ will give rise mainly to 2 lepton-jets (see right diagram in Fig. 3). Thus imposing the 4 lepton-jet cuts for process III will suppress this intermediate state and only the contribution from ZZ intermediate state will survive (see left diagram in Fig. 3). Since this ZZ contribution is very similar to the SM process I, they should have very similar cross sections after imposing 4 lepton-jet cuts as clearly seen in Table 2. On the other hand, imposing 2 lepton-jet cuts will suppress the ZZ intermediate state but keep the $\gamma_D\gamma_D$. However, the contribution of ZZ intermediate state for process III is negligible. The 2 lepton-jet cross sections of process III are several orders of magnitudes larger than the corresponding cross sections of SM process II.
- For process IV, with just basic cuts its cross section is about a factor 4 (benchmark points A and B) to 5 (benchmark points C and D) smaller than that of process III. However, due to the small mass of the dark photon (compared with Z boson mass), one can has either 4 or 2 lepton-jets in the final state. Imposing the 4 and 2 lepton-jet cuts in addition to the basic cuts for process IV have more nontrivial effects on the cross section depending on the benchmark points. For 4 lepton-jets the cross sections can reach about 3 and 1 femtobarn for benchmark points C and D respectively. For 2 lepton-jets, the cross section can reach 2 femtobarn for benchmark point A only. At these benchmark points, these cross sections are an order of magnitude larger than the corresponding cross sections of the SM process II. Other benchmark points have negligible cross sections for 4 and 2 lepton-jets as can be clearly seen in the last column of Table 2.

Table 2. Cross sections (in unit of fb) at the LHC-14 for the background processes (I and II) and dark sector processes (III and IV) with the basic cut only and basic plus 4 and 2 lepton-jet cuts at the 4 benchmark points defined in Table 1.

Cuts	Benchmark Points	I	II	III	IV
Basic	A	0.118	70.7	95.3	23.2
	B	0.118	70.7	204	45.8
	C	0.118	70.7	96.7	19.2
	D	0.118	70.7	68.3	13.1
Basic + 4 Lepton-Jets	A	9.63×10^{-3}	0.337	9.86×10^{-3}	$\leq 10^{-10}$
	B	9.63×10^{-3}	0.337	9.80×10^{-3}	$\leq 10^{-10}$
	C	9.63×10^{-3}	0.337	9.93×10^{-3}	3.05
	D	9.63×10^{-3}	0.337	9.84×10^{-3}	0.92
Basic + 2 Lepton-Jets	A	$\leq 10^{-10}$	0.08	95.3	1.75
	B	$\leq 10^{-10}$	0.08	201	$\leq 10^{-10}$
	C	$\leq 10^{-10}$	0.08	95.8	$\leq 10^{-10}$
	D	$\leq 10^{-10}$	0.08	68.2	$\leq 10^{-10}$

5. Summary
To conclude we have demonstrated that non-standard signals of multi-lepton-jets from SM Higgs decay can be used to probe the existence of light particles in the dark sector. Further studies with detector responses taking into account may therefore be worthwhile. Besides the familiar electromagnetic $U(1)$ in our visible world, perhaps such a simplest abelian symmetry may be realized in the dark sector as well.

Acknowledgments
This work was supported in part by the Ministry of Science and Technology of R.O.C. under Grant No. 101-2112-M-001-005-MY3 and the U. S. Department of Energy under Grant No. DE-SC0008541.

References
[1] G. Aad *et al.* [ATLAS Collaboration], Phys. Lett. B **716**, 1 (2012) [arXiv:1207.7214 [hep-ex]].
[2] S. Chatrchyan *et al.* [CMS Collaboration], Phys. Lett. B **716**, 30 (2012) [arXiv:1207.7235 [hep-ex]].
[3] C. F. Chang, E. Ma and T. C. Yuan, JHEP **1403**, 054 (2014) [arXiv:1308.6071 [hep-ph]].
[4] S. Gopalakrishna, S. Jung and J. D. Wells, Phys. Rev. D **78**, 055002 (2008) [arXiv:0801.3456 [hep-ph]].
[5] H. Davoudiasl, H. -S. Lee, I. Lewis and W. J. Marciano, Phys. Rev. D **88**, 015022 (2013) [arXiv:1304.4935 [hep-ph]].
[6] H. Davoudiasl, H. -S. Lee and W. J. Marciano, Phys. Rev. D **85**, 115019 (2012) [arXiv:1203.2947 [hep-ph]].
[7] A. Falkowski, J. T. Ruderman, T. Volansky and J. Zupan, Phys. Rev. Lett. **105**, 241801 (2010) [arXiv:1007.3496 [hep-ph]].
[8] A. Falkowski, J. T. Ruderman, T. Volansky and J. Zupan, JHEP **1005**, 077 (2010) [arXiv:1002.2952 [hep-ph]].
[9] D. Feldman, Z. Liu and P. Nath, Phys. Rev. D **75**, 115001 (2007) [hep-ph/0702123 [HEP-PH]].
[10] B. Holdom, Phys. Lett. B **166**, 196 (1986); Phys. Lett. B **178**, 65 (1986).
[11] K. Cheung, J. S. Lee and P. -Y. Tseng, JHEP **1305**, 134 (2013) [arXiv:1302.3794 [hep-ph]].
[12] G. Aad *et al.* [ATLAS Collaboration], New J. Phys. **15**, 043009 (2013) [arXiv:1302.4403 [hep-ex]].

A prediction of δ_{CP} for a normal neutrino mass hierarchy in an extended standard model with an A4 flavour symmetry

Dinh Nguyen Dinh[1], **Nguyen Anh Ky**[1,2], **Phi Quang Văn**[1] and **Nguyen Thi Hồng Vân**[1]

[1] Institute of physics, Vietnam academy of science and technology, 10 Dao Tan, Ba Dinh, Hanoi, Vietnam

[2] Laboratory for high energy physics and cosmology, Faculty of physics, College of science (VNU university of science), Vietnam national university, 334 Nguyen Trai, Thanh Xuan, Hanoi, Vietnam

E-mail: dndinh@iop.vast.ac.vn, anhky@iop.vast.ac.vn, pqvan@iop.vast.ac.vn, nhvan@iop.vast.ac.vn

Abstract. A method of diagonalization of a general neutrino mass matrix in an extended standard model with an A4 flavour symmetry is used. The method allows us to determine a relation between the mixing angles $(\theta_{12}, \theta_{23}, \theta_{13})$ and the CP-violation Dirac phase (δ_{CP}). The current prediction of δ_{CP} is quite good, near the 1σ region of the best fit of the experimental data at a normal neutrino mass hierarchy.

1. Introduction

The neutrino masses and mixings are among the phenomena calling for extending the standard model (SM) [1, 2]. One of the extensions of the SM is to add a flavour symmetry. A popular flavor symmetry intensively investigated in literature is A4 (see, for instance, [3, 4]) which allows to obtain a tribi-maximal neutrino mixing (TBM). The recent experimental data [5] showing, however, a non-zero mixing angle θ_{13} and a possible CP-violation Dirac phase δ_{CP}, rejects the TBM. There have been many attempts to explain these new phenomena. Here, using a perturbation method we can diagonalize a general neutrino mass matrix and derive a relation between δ_{CP} and the mixing angles θ_{ij}. The latter allows us to calculate δ_{CP} numerically using experimental data on the mixing angles.

We will make in the next section a quick introduction to the representations of A4 and their application to a simple model. In Sect. 3 we will deal with the diagonalization of the neutrino mass matrix obtained for this model from which we can get a mass spectrum and a relation between δ_{CP} and θ_{ij}.

2. Extended standard model with A4 flavour symmetry

2.1. Representations of A4 in brief

Here we give a brief information on representations of A4 [6, 7]. A4 group has 12 elements and it can be generated by two basic permutations S and T

$$S^2 = T^3 = (ST)^3 = 1. \tag{1}$$

A4 has three one-dimensional unitary representations 1, $1'$ and $1''$ generated by

$$\begin{aligned} 1 &: S = 1 \quad T = 1, \\ 1' &: S = 1 \quad T = e^{i2\pi/3} \equiv \omega, \\ 1'' &: S = 1 \quad T = e^{i4\pi/3} \equiv \omega^2, \end{aligned} \tag{2}$$

and a three-dimensional unitary representation with the generators

$$S = \begin{pmatrix} 1 & 0 & 0 \\ 0 & -1 & 0 \\ 0 & 0 & -1 \end{pmatrix}, \quad T = \begin{pmatrix} 0 & 1 & 0 \\ 0 & 0 & 1 \\ 1 & 0 & 0 \end{pmatrix}. \tag{3}$$

In applications of a group we often need to know the multiplication rule of its (irreducible) representations which in the case of A4 are

$$\begin{aligned} &1 \times 1 = 1, \\ &1' \times 1'' = 1, \\ &3 \times 3 = 1 + 1' + 1'' + 3_S + 3_{AS}. \end{aligned} \tag{4}$$

If we have two triplets $3_a \sim (a_1, a_2, a_3)$ and $3_b \sim (b_1, b_2, b_3)$, their direct product can be decomposed into irreducible representations as follows

$$\begin{aligned} 1 &= a_1 b_1 + a_2 b_2 + a_3 b_3, \\ 1' &= a_1 b_1 + \omega^2 a_2 b_2 + \omega a_3 b_3, \\ 1'' &= a_1 b_1 + \omega a_2 b_2 + \omega^2 a_3 b_3, \\ 3_1 &\sim (a_2 b_3, a_3 b_1, a_1 b_2), \\ 3_2 &\sim (a_3 b_2, a_1 b_3, a_2 b_1). \end{aligned} \tag{5}$$

The above given information is used in constructing the Lagrangian of a considered model with an A4 symmetry, for example, the Lagrangian (6).

2.2. A simple model

In any model with three active Majorana neutrinos, the most general neutrino mass matrix is symmetric and parametrized by 6 independent parameters. As an example, we consider here a model studied in [8, 9, 10]. The lepton sector (the quark sector is not considered here) of this model includes an A4 triplet N which is an $SU(2)_L$ singlet (called also right-handed neutrino), and the SM leptons among which the left-handed leptons l_L, $l = e, \mu, \tau$, transform as A4 triplets, while the right-handed ones e_R, μ_R and τ_R transform as A4 singlets 1, $1'$ and $1''$, resp. Besides the SM Higgs ϕ_h which is a singlet under A4, the scalar sector of the model has five additional

Journal of Physics: Conference Series **627** (2015) 012003 doi:10.1088/1742-6596/627/1/012003

$SU(2)_L$-singlet scalars: two A4-triplets ϕ_E, ϕ_N and three A4-singlets ξ, ξ', ξ''. The transformation rules under A4 and $SU(2)_L$ of the leptons and the scalars in this model are summarized in the table below.

	l_L	e_R	μ_R	τ_R	N	ϕ_E	ϕ_N	ξ	ξ'	ξ''	ϕ_h
A4	3	1	$1'$	$1''$	3	3	3	1	$1'$	$1''$	1
$SU(2)_L$	2	1	1	1	1	1	1	1	1	1	2

The new Yukawa term of the Lagrangian of this model is

$$-\mathcal{L}_Y^{new} = \left[\lambda_e \bar{l}_L \phi_h e_R \frac{\phi_E}{\Lambda} + \lambda_\mu \left(\bar{l}_L \phi_h\right)'' \mu_R \frac{\phi_E}{\Lambda} + \lambda_\tau \left(\bar{l}_L \phi_h\right)' \tau_R \frac{\phi_E}{\Lambda} + \lambda_N \bar{l}_L \tilde{\phi}_h N + H.c.\right]$$
$$+ c_N N^T N \phi_N + c_\xi N^T N \xi + c_{\xi'} \left(N^T N\right)'' \xi' + c_{\xi''} \left(N^T N\right)' \xi'' + H.c. \tag{6}$$

Denoting the VEVs of the scalars as

$$\langle \xi \rangle = u_a,\ \langle \xi' \rangle = u_b,\ \langle \xi'' \rangle = u_c,\ \langle \phi_E \rangle = (u_1, u_2, u_3),\ \langle \phi_N \rangle = (v_1, v_2, v_3),\ \langle \phi_h \rangle = v. \tag{7}$$

we obtain the mass matrix of the charged leptons

$$M_l = \begin{pmatrix} y_e u_1 & y_e u_2 & y_e u_3 \\ y_\mu u_1 & \omega y_\mu u_2 & \omega^2 y_\mu u_3 \\ y_\tau u_1 & \omega^2 y_\tau u_2 & \omega y_\tau u_3 \end{pmatrix} \tag{8}$$

where

$$y_e = \frac{\lambda_e}{\Lambda} v, \quad y_\mu = \frac{\lambda_\mu}{\Lambda} v, \quad y_e = \frac{\lambda_\tau}{\Lambda} v. \tag{9}$$

In the case $\langle \phi_E \rangle = (u, u, u)$, the charge lepton mass matrix can be written in the form

$$M_l = u U_0 \begin{pmatrix} y_e & 0 & 0 \\ 0 & y_\mu & 0 \\ 0 & 0 & y_\tau \end{pmatrix}, \tag{10}$$

where

$$U_0 = \frac{1}{\sqrt{3}} \begin{pmatrix} 1 & 1 & 1 \\ 1 & \omega & \omega^2 \\ 1 & \omega^2 & \omega \end{pmatrix}. \tag{11}$$

For the neutrino sector, we get a Majorana mass matrix

$$M_N = \begin{pmatrix} c_a + c_b + c_d & \epsilon_3 & \epsilon_2 \\ \epsilon_3 & c_a + \omega c_b + \omega^2 c_d & \epsilon_1 \\ \epsilon_2 & \epsilon_1 & c_a + \omega^2 c_b + \omega c_d \end{pmatrix} \equiv \begin{pmatrix} a & b & c \\ b & e & d \\ c & d & f \end{pmatrix}, \tag{12}$$

where $c_a = c_\xi u_a$, $c_b = c_{\xi'} u_b$, $c_d = c_{\xi''} u_c$, $\epsilon_1 = c_N v_1$, $\epsilon_2 = c_N v_2$, $\epsilon_3 = c_N v_3$. We also get a Dirac mass matrix

$$M_D = \begin{pmatrix} \lambda_{N_1} v & 0 & 0 \\ 0 & \lambda_{N_2} v & 0 \\ 0 & 0 & \lambda_{N_3} v \end{pmatrix} \equiv \begin{pmatrix} x & 0 & 0 \\ 0 & y & 0 \\ 0 & 0 & z \end{pmatrix}. \tag{13}$$

Following the seesaw mechanism we get a neutrino mass matrix,

$$M_\nu = -M_D^T M_N^{-1} M^D. \tag{14}$$

Journal of Physics: Conference Series **627** (2015) 012003 doi:10.1088/1742-6596/627/1/012003

which, with (12) and (13) taken into account, has the form

$$M_\nu = \frac{-1}{\det(M)} \begin{pmatrix} \left(d^2-ef\right)x^2 & (-cd+bf)xy & (-bd+ce)xz \\ (-cd+bf)xy & \left(c^2-af\right)y^2 & (-bc+ad)yz \\ (-bd+ce)xz & (-bc+ad)yz & \left(b^2-ae\right)z^2 \end{pmatrix} \equiv \begin{pmatrix} A & B & C \\ B & E & D \\ C & D & F \end{pmatrix}, \quad (15)$$

where $\det(M) = -2bcd + c^2e + b^2f + a\left(d^2-ef\right)$ is the determinant of M_N.

One of the key problems of a neutrino mass model is to diagonalize the corresponding neutrino mass matrix, in this case, the matrix (15). To solve this problem, different methods and tricks have been used. Below we will follow a perturbation approach.

3. Diagonalization of the neutrino mass matrix

In the basis of the diagonalized charged lepton mass matrix (i.e., in the basis where this matrix has a diagonal form), where $M_\nu = U_0^T M_\nu^0 U_0$, and, as shown by the current neutrino oscillation experimental data, U_{PMNS} is a small deviation from the tribi-maximal form, one always has (see also [11])

$$M_\nu = M_0 + \lambda V, \quad (16)$$

with

$$M_0 = \begin{pmatrix} A & B & -B \\ B & E & -(A-E+B) \\ -B & -(A-E+B) & E \end{pmatrix}, \quad \lambda V = \begin{pmatrix} 0 & 0 & e_1 \\ 0 & 0 & e_3 \\ e_1 & e_3 & e_2 \end{pmatrix}, \quad (17)$$

where λ is a small (perturbation) parameter and M_0 can be diagonalized

$$\mathrm{diag}(m_1^0, m_2^0, m_3^0) = U_{TBM}^\dagger M_0 U_{TBM}, \quad (18)$$

by

$$U_{TBM} = \begin{pmatrix} \sqrt{\frac{2}{3}} & \sqrt{\frac{1}{3}} & 0 \\ -\sqrt{\frac{1}{6}} & \sqrt{\frac{1}{3}} & \sqrt{\frac{1}{2}} \\ \sqrt{\frac{1}{6}} & -\sqrt{\frac{1}{3}} & \sqrt{\frac{1}{2}} \end{pmatrix} \sim \left(|1^0\rangle, |2^0\rangle, |3^0\rangle\right). \quad (19)$$

Using the perturbation expression

$$|n\rangle = |n^0\rangle + \lambda \sum_{k\neq n} |k^0\rangle \frac{V_{kn}}{m_n^0 - m_k^0}; \qquad n, k = 1, 2, 3, \quad (20)$$

with

$$V_{kn} = \langle k|V|n\rangle, \quad (21)$$

one can diagonalize the matrix M_ν by

$$U_{PMNS} = \begin{pmatrix} \sqrt{\frac{2}{3}} + \sqrt{\frac{1}{3}}x^* & \sqrt{\frac{1}{3}} - \sqrt{\frac{2}{3}}x & -\sqrt{\frac{2}{3}}y - \sqrt{\frac{1}{3}}z \\ -\sqrt{\frac{1}{6}} + \sqrt{\frac{1}{3}}x^* + \sqrt{\frac{1}{2}}y^* & \sqrt{\frac{1}{3}} + \sqrt{\frac{1}{6}}x + \sqrt{\frac{1}{2}}z^* & \sqrt{\frac{1}{2}} + \sqrt{\frac{1}{6}}y - \sqrt{\frac{1}{3}}z \\ \sqrt{\frac{1}{6}} - \sqrt{\frac{1}{3}}x^* + \sqrt{\frac{1}{2}}y^* & -\sqrt{\frac{1}{3}} - \sqrt{\frac{1}{6}}x + \sqrt{\frac{1}{2}}z^* & \sqrt{\frac{1}{2}} - \sqrt{\frac{1}{6}}y + \sqrt{\frac{1}{3}}z \end{pmatrix}, \quad (22)$$

where

$$x = \frac{\sqrt{2}}{6}\left(\frac{2e_3 - e_1 - e_2}{m_1^0 - m_2^0}\right), \quad y = \frac{\sqrt{3}}{6}\left(\frac{2e_1 + e_2}{m_1^0 - m_3^0}\right), \quad z = \frac{1}{\sqrt{6}}\left(\frac{e_1 + e_2}{m_2^0 - m_3^0}\right). \quad (23)$$

From (22) a relation between the mixing angles and CP-violation Dirac phase δ_{CP} can be obtained and briefly discussed below. The mass spectrum derived by the present method of diagonalization of the mass matrix M_ν will be discussed in a separate work because it needs more analysis on the VEV's of the scalars and the Yukawa coupling coefficients as well as different phenomena and experimental results.

4. Relation between mixing angles and δ_{CP}

Denoting with U_{ij}, $i,j = 1,2,3$, a matrix element of (22), we obtain the relation

$$U_{21} + \sqrt{2}U_{22} - U_{31} - \sqrt{2}U_{32} = 2U_{11} - \sqrt{2}U_{12}. \tag{24}$$

The latter compared with the elements of the matrix U_{PMNS} of a general form

$$U_{PMNS} = \begin{pmatrix} c_{12}c_{13} & s_{12}c_{13} & s_{13}e^{-i\delta} \\ -c_{23}s_{12} - s_{13}s_{23}c_{12}e^{i\delta} & c_{23}c_{12} - s_{13}s_{23}s_{12}e^{i\delta} & s_{23}c_{13} \\ s_{23}s_{12} - s_{13}c_{23}c_{12}e^{i\delta} & -s_{23}c_{12} - s_{13}c_{23}s_{12}e^{i\delta} & c_{23}c_{13}, \end{pmatrix}, \tag{25}$$

where $s_{ij} = \sin\theta_{ij}$, $c_{ij} = \cos\theta_{ij}$; $i,j = 1,2,3$, leads to a relation between $\delta_{CP} \equiv \delta$ and the mixing angles θ_{ij}:

$$\cos\delta\tan\theta_{13} = \frac{(\sqrt{2} - \tan\theta_{12})}{(1+\sqrt{2}\tan\theta_{12})(1-\tan\theta_{23})}\left(\frac{\sqrt{2}}{c_{23}} - \frac{1+\tan\theta_{23}}{c_{13}}\right). \tag{26}$$

In general, it is not easy to determine δ_{CP} both theoretically and experimentally. Here, we obtain an explicit δ_{CP} as a function of the mixing angles which, thus, δ_{CP}, could be determined experimentally. Based on the relation (1) we can calculate δ_{CP} numerically using experimental inputs. The distributions of δ_{CP} are plotted in Fig. 1 and Fig. 2.

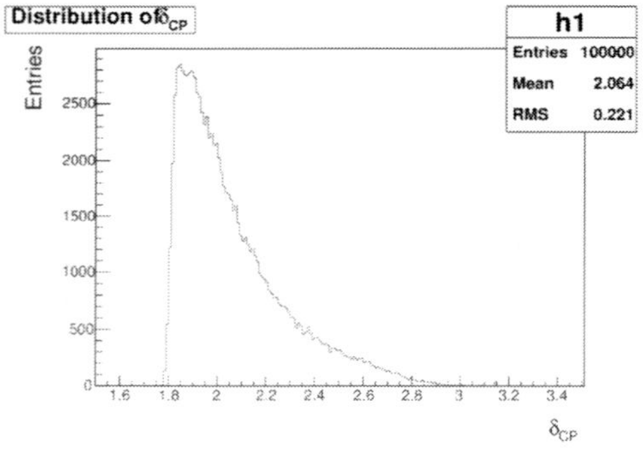

Figure 1: Distribution of δ_{CP}

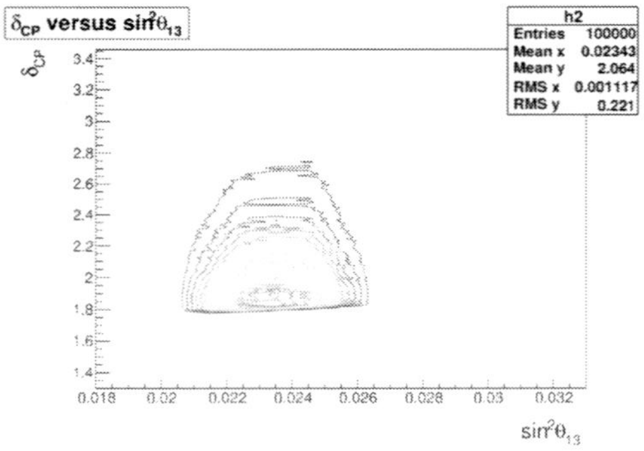

Figure 2: δ_{CP} versus $\sin^2\theta_{13}$.

From these figures we see that δ_{CP} distributes in the region $1.78 < \delta_{CP} < 3.2$. This distribution has a mean value at $\delta_{CP} = 2.064$ which is on the edge of the 1σ region from the best fit value (BFV) to the experimental data and reaches a maximum value around $\delta_{CP} = 1.9$ which is between 1σ from the BFV [5].

Conclusions

A general neutrino mass matrix in an extended SM with an A4 flavour symmetry can be diagonalized by a pertubation method. Then, an explicit dependence between the CP-violation Dirac phase and the mixing angles leads to a quite good fit (near the 1σ region of the best fit) with recent experimental data at the normal hierarchy [5]. This result provides a more explicit form of neutrino mass and mixing matrices. The case with the inverse hierarchy is being investigated.

We are also considering pertubation of higher orders which may give a better fit (e.g., between the 1σ region) with the experimental data.

Acknowledgements: This work is supported by Vietnam's National Foundation for Science and Technology Development (NAFOSTED) under the grant No 103.03-2012.49.

References

[1] Bilenky S 2010 *Introduction to the physics of massive and mixed neutrinos* (Berlin: Springer).
[2] Giunti C, Kim C W 2007 *Fundamentals of neutrino physics and astrophysics* (New York: Oxford university Press).
[3] Altarelli G, Feruglio F and Merlo L 2013 *Fortsch. Phys.* **61**, 507.
[4] Ferreira P M, Lavoura L and Ludl P O 2013 *Phys. Lett.* B **726**, 767.
[5] Capozzi F, Fogli G L, Lisi E, Marrone A, Montanino D and Palazzo A 2014 *Phys. Rev.* D **89**, 093018.
[6] Altarelli G, Feruglio F, Merlo L and Stamou E 2012 *JHEP* **1208**, 021.
[7] King S F and Luhn C 2013 *Rep. Prog. Phys.* **76**, 056201.
[8] Ishimori H and Ma E 2012 *Phys. Rev.* D **86**, 045030.
[9] Ma E, Natale A and Rashed A 2012 *Int. J. Mod. Phys.* A **27**, 1250134.
[10] Honda M and Tanimoto M 2008 *Prog. Theor. Phys.* **119**, 583.
[11] Brahmachari B and Raychaudhuri A 2012 *Phys. Rev.* D **86**, 051302.

Journal of Physics: Conference Series **627** (2015) 012004 doi:10.1088/1742-6596/627/1/012004

Enhanced Electron EDM with Minimal Flavor Violation

Jusak Tandean

Department of Physics and Center for Theoretical Sciences, National Taiwan University, Taipei 106, Taiwan

E-mail: jtandean@yahoo.com

Abstract. The latest data from the ACME experiment have led to the most stringent limit to date on the electric dipole moment d_e of the electron. Nevertheless, the standard model (SM) prediction for d_e is many orders of magnitude below the new result, making this observable a powerful probe for physics beyond the SM. We perform a model-independent study of d_e in the SM plus right handed neutrinos and its extension with the seesaw mechanism under the framework of minimal flavor violation (MFV). We find that d_e crucially depends on whether neutrinos are Dirac or Majorana fermions. In the Majorana case, d_e can reach its measured bound, which therefore constrains the scale of MFV to be above a few hundred GeV. We also consider extra CP-violating sources in the Yukawa couplings of the right-handed neutrinos. Such new sources can have important effects on d_e.

1. Introduction

Electric dipole moments (EDMs) constitute highly sensitive indicators for the presence of new sources of the violation of charge parity (CP) and time reversal (T) symmetries beyond the standard model (SM) of particle physics [1, 2, 3, 4]. Recently the ACME experiment [5], which searched for the electron EDM, d_e, utilizing the polar molecule thorium monoxide, has reported a fresh result of $d_e = (-2.1 \pm 3.7_{\rm stat} \pm 2.5_{\rm syst}) \times 10^{-29}\, e\,{\rm cm}$, which corresponds to an upper bound of $|d_e| < 8.7 \times 10^{-29}\, e\,{\rm cm}$ at 90% confidence level. This is more stringent than the previous best limit by about an order of magnitude, but still way above the SM expectation for d_e, which is at the level of $10^{-44}\, e\,{\rm cm}$ [6]. Hence there is plenty of room between the current bound on d_e and its SM value where potential new physics may be detected in future measurements.

Extra ingredients beyond the SM can boost d_e considerably with respect to its SM prediction, even up to its experimental limit. Such tremendous enhancement of d_e may hail from various origins depending on the specifics of the new physics models. Therefore it is desirable to carry out an analysis of d_e beyond the SM which deals with some general features of the physics in a model-independent fashion. This turns out to be feasible under the framework of the so-called minimal flavor violation (MFV) which presupposes that the sources of all flavor-changing neutral currents (FCNC) and CP violation reside in renormalizable Yukawa couplings [7, 8]. This offers a systematic method to organize and study possible SM-related flavor- and CP-violating new interactions.

Here we discuss an MFV treatment of d_e recently performed in Refs. [9, 10] within the SM with three right-handed neutrinos and its extension with the neutrino seesaw mechanism. The relevant contributions arise from effective dipole operators. As will be shown below, the predicted

Journal of Physics: Conference Series **627** (2015) 012004 doi:10.1088/1742-6596/627/1/012004

size of d_e depends significantly on whether light neutrinos are Dirac or Majorana in nature. In the Majorana case, d_e can be as large as its experimental bound, which therefore limits the scale of MFV to be above a few hundred GeV or higher [9, 10].

2. Leptonic MFV

In the SM slightly expanded with the addition of three right-handed neutrinos, the renormalizable Lagrangian for lepton masses is given by

$$\mathcal{L}_{\rm m} \;=\; -(Y_\nu)_{kl}\,\bar{L}_{k,L}\,\nu_{l,R}\,\tilde{H} \;-\; (Y_e)_{kl}\,\bar{L}_{k,L}\,E_{l,R}\,H \;-\; \tfrac{1}{2}(M_\nu)_{kl}\,\overline{\nu^{\rm c}_{k,R}}\,\nu_{l,R} \;+\; {\rm H.c.}\;, \tag{1}$$

where $k,l = 1,2,3$ are summed over, $L_{k,L}$ represents left-handed lepton doublets, $\nu_{l,R}$ ($E_{l,R}$) denotes right-handed neutrinos (charged leptons), $Y_{\nu,e}$ are matrices for the Yukawa couplings, H is the Higgs doublet, $\tilde{H} = i\tau_2 H^*$, and M_ν is the Majorana mass matrix for $\nu_{l,R}$. The M_ν part is essential for the seesaw mechanism to generate light neutrino masses [11].

Since it is still unknown whether light neutrinos are Dirac or Majorana particles, we deal with the two possibilities separately. If neutrinos are of Dirac nature, the M_ν terms in Eq. (1) are absent, and according to the MFV hypothesis for leptons [8] the Lagrangian is formally invariant under the global group ${\rm U}(3)_L \times {\rm U}(3)_\nu \times {\rm U}(3)_E = G_\ell \times {\rm U}(1)_L \times {\rm U}(1)_\nu \times {\rm U}(1)_E$ with $G_\ell = {\rm SU}(3)_L \times {\rm SU}(3)_\nu \times {\rm SU}(3)_E$. This entails that the three generations of $L_{k,L}$, $\nu_{k,R}$, and $E_{k,R}$ transform as fundamental representations of ${\rm SU}(3)_{L,\nu,E}$, respectively,

$$L_L \to V_L L_L\;, \qquad \nu_R \to V_\nu \nu_R\;, \qquad E_R \to V_E E_R\;, \qquad V_{L,\nu,E} \in {\rm SU}(3)\;, \tag{2}$$

whereas the Yukawa couplings are spurions transforming according to

$$Y_\nu \to V_L Y_\nu V_\nu^\dagger\;, \qquad Y_e \to V_L Y_e V_E^\dagger\;. \tag{3}$$

Taking advantage of the invariance under G_ℓ, we work in the basis where

$$Y_e \;=\; \frac{\sqrt{2}}{v}\,{\rm diag}(m_e, m_\mu, m_\tau)\;, \tag{4}$$

with $v \simeq 246\,{\rm GeV}$ being the vacuum expectation value of H, and the fields $\nu_{k,L}$, $\nu_{k,R}$, $E_{k,L}$, and $E_{k,R}$ refer to the mass eigenstates. We can then express $L_{k,L}$ and Y_ν in terms of the Pontecorvo-Maki-Nakagawa-Sakata (PMNS) neutrino mixing matrix $U_{\rm PMNS}$ as

$$L_{k,L} = \begin{pmatrix} (U_{\rm PMNS})_{kl}\,\nu_{l,L} \\ E_{k,L} \end{pmatrix}, \qquad Y_\nu \;=\; \frac{\sqrt{2}}{v}\,U_{\rm PMNS}\,\hat{m}_\nu\;, \qquad \hat{m}_\nu \;=\; {\rm diag}(m_1, m_2, m_3)\;, \tag{5}$$

where $m_{1,2,3}$ are the light neutrino eigenmasses and in the standard parametrization [12]

$$U_{\rm PMNS} = \begin{pmatrix} c_{12}\,c_{13} & s_{12}\,c_{13} & s_{13}\,e^{-i\delta} \\ -s_{12}\,c_{23} - c_{12}\,s_{23}\,s_{13}\,e^{i\delta} & c_{12}\,c_{23} - s_{12}\,s_{23}\,s_{13}\,e^{i\delta} & s_{23}\,c_{13} \\ s_{12}\,s_{23} - c_{12}\,c_{23}\,s_{13}\,e^{i\delta} & -c_{12}\,s_{23} - s_{12}\,c_{23}\,s_{13}\,e^{i\delta} & c_{23}\,c_{13} \end{pmatrix}, \tag{6}$$

with δ being the CP violation phase, $c_{kl} = \cos\theta_{kl}$, and $s_{kl} = \sin\theta_{kl}$.

If neutrinos are of Majorana nature, the M_ν part in Eq. (1) is allowed. As a consequence, for $M_\nu \gg M_{\rm D} = vY_\nu/\sqrt{2}$ the seesaw mechanism [11] becomes operational involving the $6{\times}6$ neutrino mass matrix

$$\mathsf{M} \;=\; \begin{pmatrix} 0 & M_{\rm D} \\ M_{\rm D}^{\rm T} & M_\nu \end{pmatrix} \tag{7}$$

in the $(U^*_{\rm PMNS}\nu^{\rm c}_L, \nu_R)^{\rm T}$ basis. The resulting mass-matrix for the light neutrinos is

$$m_\nu \;=\; -\frac{v^2}{2}\, Y_\nu M_\nu^{-1} Y_\nu^{\rm T} \;=\; U_{\rm PMNS}\, \hat{m}_\nu\, U^{\rm T}_{\rm PMNS} \,, \tag{8}$$

where now $U_{\rm PMNS}$ contains the diagonal matrix $P = {\rm diag}(e^{i\alpha_1/2}, e^{i\alpha_2/2}, 1)$ multiplied from the right, $\alpha_{1,2}$ being the Majorana phases. It follows that Y_ν in Eq. (5) is no longer valid, and one can instead take Y_ν to be [13]

$$Y_\nu \;=\; \frac{i\sqrt{2}}{v}\, U_{\rm PMNS}\, \hat{m}_\nu^{1/2} O M_\nu^{1/2} \,, \tag{9}$$

where O is a matrix satisfying $OO^{\rm T} = \mathbb{1}$ and $M_\nu = {\rm diag}(M_1, M_2, M_3)$. As we see later, O can give rise to new CP-violation effects besides those from $U_{\rm PMNS}$.

To arrange nontrivial two-lepton FCNC and CP-violating interactions satisfying the MFV principle, one assembles an arbitrary number of the Yukawa coupling matrices $Y_\nu \sim (3, \bar{3}, 1)$ and $Y_e \sim (3, 1, \bar{3})$ as well as their Hermitian conjugates to devise the G_ℓ representations $\Delta_{\ell 8} \sim (8, 1, 1)$, $\Delta_{\nu 8} \sim (1, 8, 1)$, $\Delta_{e8} \sim (1, 1, 8)$, $\Delta_\nu \sim (\bar{3}, 3, 1)$, and $\Delta_e \sim (\bar{3}, 1, 3)$, combines them with two lepton fields to arrive at the G_ℓ-invariant objects $\bar{L}_L \Delta_{\ell 8} L_L$, $\bar{\nu}_R \Delta_{\nu 8} \nu_R$, $\bar{E}_R \Delta_{e8} E_R$, $\bar{\nu}_R \Delta_\nu L_L$, and $\bar{E}_R \Delta_e L_L$, attaches appropriate numbers of the Higgs field H and SM gauge fields to form singlets under the SM gauge group, and also contracts all Lorentz indices. Since fermion EDMs flip chirality, only the last combination, $\bar{E}_R \Delta_e L_L$, pertains to our examination of d_e. We can choose to write $\Delta_e = Y_e^\dagger \Delta$ with Δ consisting of products of $\mathsf{A} = Y_\nu Y_\nu^\dagger$ and $\mathsf{B} = Y_e Y_e^\dagger$.

Formally Δ comprises an infinite number of terms, $\Delta = \sum \xi_{ijk\cdots} \mathsf{A}^i \mathsf{B}^j \mathsf{A}^k \cdots$. The MFV hypothesis implies that the coefficients $\xi_{ijk\cdots}$ have to be real because otherwise they would introduce new CP-violating sources beyond what is already contained in $Y_{e,\nu}$. Using the Cayley-Hamilton identity $X^3 = X^2\, {\rm Tr}X + X[{\rm Tr}X^2 - ({\rm Tr}X)^2]/2 + \mathbb{1}\, {\rm Det}X$ for a 3×3 invertible matrix X, this infinite series can be resummed into a finite number of terms [14]:

$$\begin{aligned}\Delta \;=\;& \xi_1 \mathbb{1} + \xi_2 \mathsf{A} + \xi_3 \mathsf{B} + \xi_4 \mathsf{A}^2 + \xi_5 \mathsf{B}^2 + \xi_6 \mathsf{AB} + \xi_7 \mathsf{BA} + \xi_8 \mathsf{ABA} + \xi_9 \mathsf{BA}^2 + \xi_{10} \mathsf{BAB} + \xi_{11} \mathsf{AB}^2 \\ &+ \xi_{12} \mathsf{ABA}^2 + \xi_{13} \mathsf{A}^2\mathsf{B}^2 + \xi_{14} \mathsf{B}^2\mathsf{A}^2 + \xi_{15} \mathsf{B}^2\mathsf{AB} + \xi_{16} \mathsf{AB}^2\mathsf{A}^2 + \xi_{17} \mathsf{B}^2\mathsf{A}^2\mathsf{B} \,,\end{aligned} \tag{10}$$

where $\mathbb{1}$ is a 3×3 unit matrix. Although $\xi_{ijk\cdots}$ are real, the reduction of the infinite series into the 17 terms can make the coefficients ξ_r in Eq. (10) complex due to imaginary parts among the traces of the matrix products $\mathsf{A}^i \mathsf{B}^j \mathsf{A}^k \cdots$. Such imaginary contributions turn out to be proportional to the Jarlskog invariant ${\rm Im\,Tr}(\mathsf{A}^2\mathsf{BAB}^2) = (i/2)\,{\rm Det}[\mathsf{A}, \mathsf{B}] \ll 1$ [15]. The implication is that, with all $\xi_{ijk\cdots}$ being expected to be at most of $\mathcal{O}(1)$, the impact of ξ_r on d_e is suppressed by a factor of $m_\mu^2 m_\tau^2/v^4$ compared to the contribution from ABA^2 which has the smallest number of suppressive factor Y_e among the products in Eq. (10) that contribute to d_e. Therefore, hereafter we ignore ${\rm Im}\,\xi_r$.

3. Lepton EDMs with MFV

The EDM d_l of a lepton l is described by $\mathcal{L}_d = -(id_l/2)\bar{l}\sigma^{\kappa\omega}\gamma_5 l F_{\kappa\omega}$, where $F_{\kappa\omega}$ is the photon field strength tensor. In the MFV framework, this arises at lowest order from the operators in the effective Lagrangian [8]

$$\mathcal{L}_{\rm eff} \;=\; \frac{1}{\Lambda^2}\Big(g' \bar{E}_R Y_e^\dagger \Delta_1 \sigma_{\kappa\omega} H^\dagger L_L B^{\kappa\omega} \;+\; g \bar{E}_R Y_e^\dagger \Delta_2 \sigma_{\kappa\omega} H^\dagger \tau_j L_L W_j^{\kappa\omega} \Big) \;+\; {\rm H.c.}\,, \tag{11}$$

where mass scale Λ characterizes the heavy new physics underlying these flavor-violating interactions, W and B denote the SM gauge fields with coupling constants g and g', respectively, τ_j are Pauli matrices, and $\Delta_{1,2}$ are of the form in Eq. (10) with generally different ξ_r's.

Expanding Eq. (11), we find that the charged-lepton EDMs are proportional to $\mathrm{Im}(Y_e^\dagger \Delta_{1,2})_{kk}$, but that in $\Delta_{1,2}$ only the ABA^2 and $\mathsf{AB}^2\mathsf{A}^2$ terms are relevant. Thus for the electron

$$d_e = \frac{\sqrt{2}\,ev}{\Lambda^2}\Big[\xi_{12}^\ell\,\mathrm{Im}(Y_e^\dagger \mathsf{ABA}^2)_{11} + \xi_{16}^\ell\,\mathrm{Im}(Y_e^\dagger \mathsf{AB}^2\mathsf{A}^2)_{11}\Big]\,, \tag{12}$$

where $\xi_r^\ell = \xi_r^{(1)} - \xi_r^{(2)}$. If neutrinos are Dirac particles, we obtain from Eqs. (5) and (12)

$$d_e^{\rm D} = \frac{32em_e}{\Lambda^2}\left[\xi_{12}^\ell + \frac{2(m_\mu^2+m_\tau^2)}{v^2}\,\xi_{16}^\ell\right]\frac{(m_\mu^2-m_\tau^2)(m_1^2-m_2^2)(m_2^2-m_3^2)(m_3^2-m_1^2)}{v^8}\,J_\ell\,, \tag{13}$$

invariant for $U_{\rm PMNS}$. In the Majorana neutrino case, if the right-handed neutrinos $\nu_{k,R}$ are degenerate, $M_\nu = \mathcal{M}\mathbb{1}$, and O is a real orthogonal matrix, from Eq. (9) we have

$$\mathsf{A} = \frac{2}{v^2}\,\mathcal{M}\,U_{\rm PMNS}\,\hat{m}_\nu U_{\rm PMNS}^\dagger \tag{14}$$

and consequently

$$d_e^{\rm M} = \frac{32e\mathcal{M}^3 m_e}{\Lambda^2 v^8}(m_\mu^2-m_\tau^2)(m_1-m_2)(m_2-m_3)(m_3-m_1)\,\xi_{12}^\ell\,J_\ell\,, \tag{15}$$

neglecting the ξ_{16}^ℓ term. Since $m_k \ll \mathcal{M}$, obviously $d_e^{\rm D}$ is highly suppressed relative to $d_e^{\rm M}$.

In the preceding paragraph, d_e depends on the CP-violating Dirac phase δ in $U_{\rm PMNS}$, and the Majorana phases $\alpha_{1,2}$ therein do not enter. However, if $\nu_{k,R}$ are not degenerate, nonzero $\alpha_{1,2}$ can lead to an extra effect on d_e even with a real $O \neq \mathbb{1}$. If O is complex, its phases may induce an additional contribution to d_e, whether or not $\nu_{k,R}$ are degenerate. We explore these scenarios numerically later.

To evaluate d_e, we need the values of the various pertinent quantities, such as the elements of $U_{\rm PMNS}$ as well as the masses of neutrinos and charged leptons. In Table 1, we have listed $\sin^2\theta_{kl}$ and δ from a recent fit to global neutrino data [16]. Most of these numbers depend on whether neutrino masses fall into a normal hierarchy (NH) or an inverted one (IH). Since the Majorana phases $\alpha_{1,2}$ remain undetermined, we will select specific values for them in our illustrations.

For Dirac neutrinos, we scan the empirical ranges of the parameters from Table 1 to maximize $d_e^{\rm D}$ in Eq. (13). The result in the NH or IH case is $d_e^{\rm D} = 1.3\times10^{-99}\,\xi_{12}^\ell\,({\rm GeV}^2/\Lambda^2)\,e\,{\rm cm}$. This is negligible compared to the latest experimental upper bound [5].

Table 1. Results of a recent fit to the global data on neutrino oscillations [16]. The neutrino mass hierarchy may be normal ($m_1 < m_2 < m_3$) or inverted ($m_3 < m_1 < m_2$).

Observable	NH	IH
$\sin^2\theta_{12}$	0.308 ± 0.017	0.308 ± 0.017
$\sin^2\theta_{23}$	$0.425^{+0.029}_{-0.027}$	$0.437^{+0.059}_{-0.029}$
$\sin^2\theta_{13}$	$0.0234^{+0.0022}_{-0.0018}$	0.0239 ± 0.0021
δ/π	$1.39^{+0.33}_{-0.27}$	$1.35^{+0.24}_{-0.39}$
$\delta m^2 = m_2^2 - m_1^2$	$(7.54^{+0.26}_{-0.22})\times10^{-5}\ {\rm eV}^2$	$(7.54^{+0.26}_{-0.22})\times10^{-5}\ {\rm eV}^2$
$\Delta m^2 = \lvert m_3^2 - (m_1^2+m_2^2)/2\rvert$	$(2.44^{+0.08}_{-0.06})\times10^{-3}\ {\rm eV}^2$	$(2.40 \pm 0.07)\times10^{-3}\ {\rm eV}^2$

If neutrinos are Majorana fermions, in contrast, d_e can be sizable. We start with the simplest possibility that $\nu_{k,R}$ are degenerate, $M_\nu = \mathcal{M}\mathbb{1}$, and the O matrix in Eq. (9) is real. Thus d_e is already given in Eq. (15). Scanning again the empirical parameter ranges in Table 1 to maximize $d_e^{\rm M}$, we then obtain for $m_1 = 0$ $(m_3 = 0)$ in the NH (IH) case

$$\frac{d_e^{\rm M}}{e\,{\rm cm}} = 47\,(5.2)\times 10^{-24}\left(\frac{\mathcal{M}}{10^{15}\,{\rm GeV}}\right)^3\left(\frac{\rm GeV}{\hat\Lambda}\right)^2, \qquad \hat\Lambda = \frac{\Lambda}{|\xi_{12}^\ell|^{1/2}}, \tag{16}$$

where $\mathcal{M}$ is specified below. Hence $|d_e|_{\rm exp} < 8.7\times 10^{-29}\,e\,{\rm cm}$ [5] translates into

$$\hat\Lambda > 0.74\,(0.24)\,{\rm TeV}\left(\frac{\mathcal{M}}{10^{15}\,{\rm GeV}}\right)^{3/2}. \tag{17}$$

Since $d_e^{\rm M}$ in Eq. (16) is proportional to $\mathcal{M}^3$, one might naively think that $d_e^{\rm M}$ can easily reach its measured bound, which would therefore constrain $\hat\Lambda$ to an arbitrarily high level with a very large $\mathcal{M}$. However, the size of $\mathcal{M}$ is capped based on the condition that the series in Eq. (10) which supposedly includes arbitrarily high powers of A and B must converge [9, 15]. Therefore, in this work, we require that the biggest eigenvalue of A not exceed 1. Accordingly, we arrive at $\mathcal{M} = 6.16\,(6.22)\times 10^{14}\,{\rm GeV}$ in Eq. (16) for the NH (IH) case, roughly similar to the expected seesaw scales in some grand unified theories. Hence $\hat\Lambda > 360\,(120)\,{\rm GeV}$ from Eq. (17). For comparison with later examples, these $\mathcal{M}$ numbers imply $d_e^{\rm M}\hat\Lambda^2 = 1.1\,(0.13)\times 10^{-23}\,e\,{\rm cm}$.

With $\nu_{k,R}$ again degenerate, $M_\nu = \mathcal{M}\mathbb{1}$, but O complex, we have in Eq. (12)

$$\mathsf{A} = \frac{2}{v^2}\mathcal{M}\,U_{\rm PMNS}\,\hat m_\nu^{1/2} O O^\dagger \hat m_\nu^{1/2} U_{\rm PMNS}^\dagger \tag{18}$$

In general $OO^\dagger = e^{2i\mathsf{R}}$, where R is a real antisymmetric matrix with nonzero elements denoted by $r_1 = \mathsf{R}_{12} = -\mathsf{R}_{21}$, $r_2 = \mathsf{R}_{13} = -\mathsf{R}_{31}$, and $r_3 = \mathsf{R}_{23} = -\mathsf{R}_{32}$. Since $OO^\dagger$ is not diagonal, the Majorana phases in $U_{\rm PMNS}$ can also enter A if $\alpha_{1,2} \neq 0$. We concentrate first on the CP-violating effect of O by setting $\alpha_{1,2} = 0$. For illustrations, we pick two possible sets of $r_{1,2,3}$, namely, (i) $r_1 = -r_2 = r_3 = -\rho$ and (ii) $r_1 = 2r_2 = 3r_3 = \rho$, and employ the central values of the data in Table 1. In Figure 1 we present the resulting $d_e^{\rm M}\hat\Lambda^2$ versus ρ for the NH (IH) of light neutrino masses with $m_{1(3)} = 0$. Since δ is not yet well-determined, we also depict the variations of $d_e^{\rm M}$ over the one-sigma ranges of δ quoted in Table 1 with the lighter blue and red bands. Within these bands, the blue and red solid curves belong, respectively, to the NH and IH central values in the table. We also graph the (dashed) curves for $\delta = 0$ to reveal the CP-violating role of O alone.

With $\alpha_{1,2} = 0$, the CP-violating effect of O can still occur even if it is real, provided that $\nu_{k,R}$ are nondegenerate, in which case

$$\mathsf{A} = \frac{2}{v^2}U_{\rm PMNS}\,\hat m_\nu^{1/2} O M_\nu O^\dagger \hat m_\nu^{1/2} U_{\rm PMNS}^\dagger \tag{19}$$

from Eq. (9). For instance, assuming that O is real, $O = e^{\mathsf{R}}$ with (a) $r_1 = -r_2 = r_3 = -\rho$ and (b) $r_1 = 2r_2 = 3r_3 = \rho$, and that $M_\nu = \mathcal{M}\,{\rm diag}(1, 0.8, 1.2)$, we show the resulting $d_e^{\rm M}\hat\Lambda^2$ versus ρ in Figure 2, where only the $\delta \neq 0$ curves are nonvanishing and the sinusoidal behavior of d_e is visible. As in the previous figure, we also display the variations of $d_e^{\rm M}$ over the one-sigma ranges of δ from Table 1.

All of these results in Figures 1 and 2 clearly demonstrate that O can generate potentially significant extra effects of CP violation which can exceed those of δ. The latter point is most

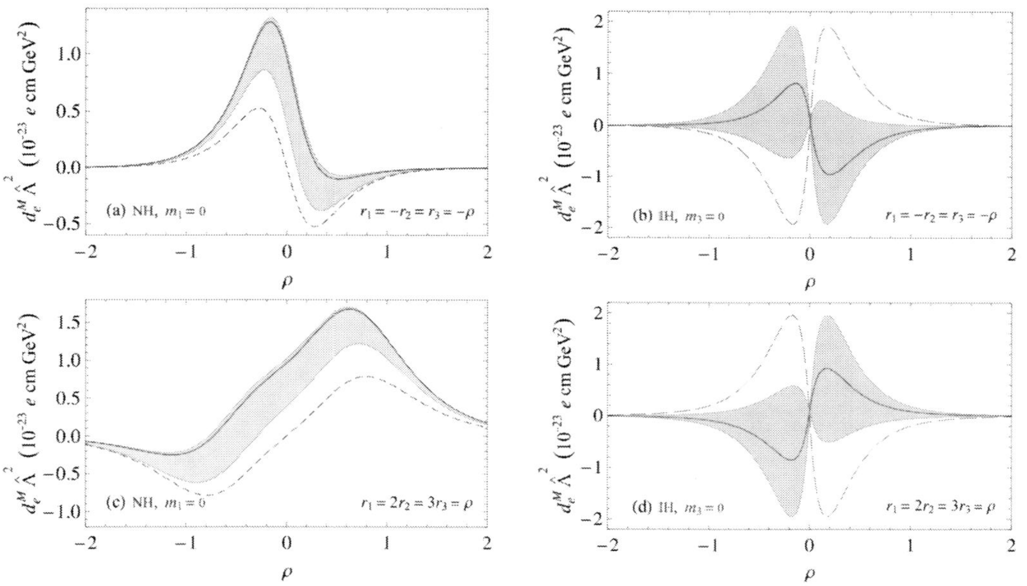

Figure 1. Dependence of $d_e^{\rm M}$ times $\hat{\Lambda}^2 = \Lambda^2/|\xi_{12}^{\ell}|$ on the O-matrix parameter ρ for zero Majorana phases, $\alpha_{1,2} = 0$, degenerate $\nu_{k,R}$, and complex O with (a,b) $r_1 = -r_2 = r_3 = -\rho$ and (c,d) $r_1 = 2r_2 = 3r_3 = \rho$, as discussed in the text. The lighter blue and red bands reflect the one-sigma ranges of δ, while the solid and dashed curves correspond, respectively, to its central values and to $\delta = 0$.

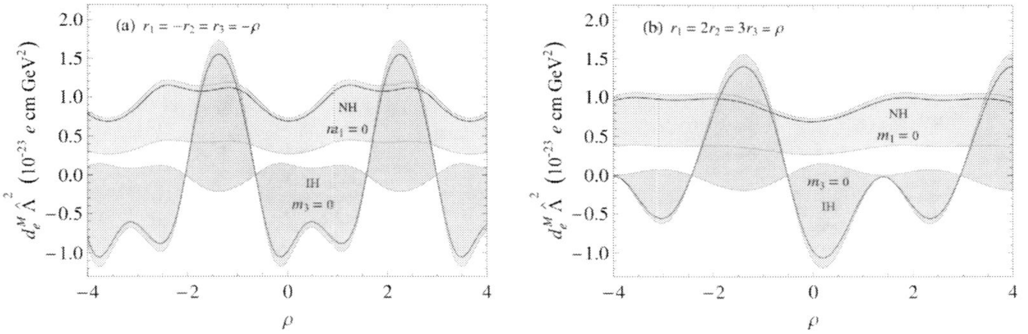

Figure 2. Dependence of $d_e^{\rm M}\hat{\Lambda}^2$ on ρ for zero Majorana phases, $\alpha_{1,2} = 0$, nondegenerate $\nu_{k,R}$ with $M_\nu = \mathcal{M}\,{\rm diag}(1, 0.8, 1.2)$, and real $O = e^{\rm R}$ with (a) $r_1 = -r_2 = r_3 = -\rho$ and (b) $r_1 = 2r_2 = 3r_3 = \rho$, as discussed in the text. The lighter blue and red bands reflect the one-sigma ranges of δ, while the solid curves correspond to its central values.

noticeable in Figure 1(b,d) from comparing the IH $\delta \neq 0$ regions at $\rho \sim 0$ with the extreme values of the corresponding IH $\delta = 0$ curves.

To see the effect of the Majorana phases, we entertain a couple of possibilities: (i) $M_\nu = \mathcal{M}\mathbb{1}$ and $O = e^{i\rm R}$ and (ii) $M_\nu = \mathcal{M}\,{\rm diag}(1, 0.8, 1.2)$ and $O = e^{\rm R}$, both with the two sets of $r_{1,2,3}$ chosen earlier. Fixing $\alpha_1 = 0$ and $\rho = 1/2$, we depict the resulting dependence of $d_e^{\rm M}$ on α_2 in Figures 3 and 4 for nonzero δ within its one-sigma ranges from Table 1 and also for $\delta = 0$. It is

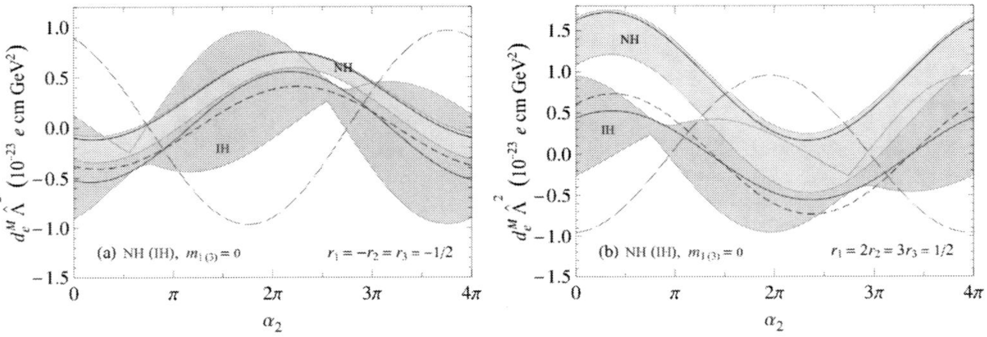

Figure 3. Dependence of $d_e^{\mathrm{M}}\hat{\Lambda}^2$ on α_2 for $\alpha_1 = 0$, degenerate $\nu_{k,R}$, and $O = e^{i\mathrm{R}}$ with (a) $r_{1,3} = -r_2 = -1/2$ and (b) $r_1 = 2r_2 = 3r_3 = 1/2$, as explained in the text. The bands and curves have the same meanings as in preceding figures.

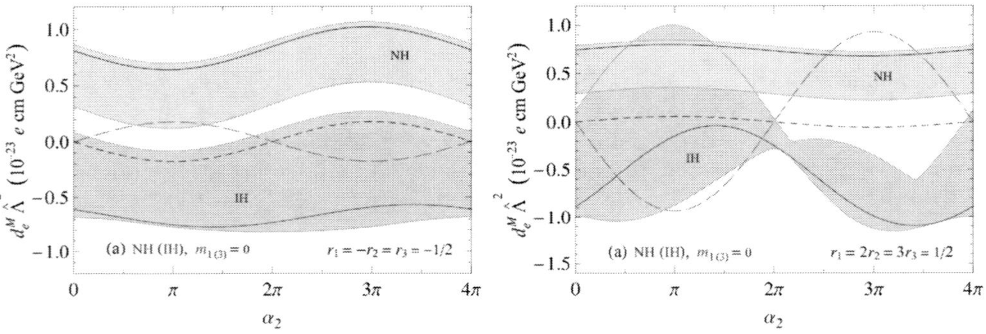

Figure 4. The same as Figure 3, except $\nu_{k,R}$ are nondegenerate with $M_\nu = \mathcal{M}\,\mathrm{diag}(1, 0.8, 1.2)$ and $O = e^{\mathrm{R}}$.

evident from these graphs that the Majorana phases produce additional important CP-violating impact on d_e beyond δ.

For the EDMs of the muon and tau lepton, one could do a similar analysis. However, their experimental limits are still much weaker than $|d_e|_{\mathrm{exp}}$. Consequently, they lead to bounds on $\hat{\Lambda}$ which are not competitive to that in the electron case [9, 10].

4. Conclusions

We have investigated the electron EDM, d_e, under the MFV framework and found that d_e can reach its experimental limit if neutrinos are Majorana in nature. Moreover, from the latest data on d_e, we infer that the MFV scale has to be a few hundred GeV or higher. We demonstrate that d_e has the potential to probe not only the Dirac phase in the lepton mixing matrix, but also the Majorana phases therein, as well as extra CP-violation sources in the Yukawa couplings of the right-handed neutrinos.

Acknowledgments

This work was supported in part by the MOE Academic Excellence Program (Grant No. 102R891505) and the NCTS.

References

[1] He X G, McKellar B H J, and Pakvasa S 1989 *Int. J. Mod. Phys.* A **4** 5011
[2] Bernreuther W and Suzuki M 1991 *Rev. Mod. Phys.* **63** 313
[3] Fukuyama T 2012 *Int. J. Mod. Phys.* A **27** 1230015
[4] Engel J, Ramsey-Musolf M J, and van Kolck U 2013 *Prog. Part. Nucl. Phys.* **71** 21
[5] Baron J *et al* 2014 *Science* **343** 269
[6] Pospelov M and Ritz A 2014 *Phys. Rev.* D **89** 056006
[7] D'Ambrosio G, Giudice G F, Isidori G, and Strumia A 2002 *Nucl. Phys.* B **645** 155
[8] Cirigliano V, Grinstein B, Isidori G, and Wise M B 2005 *Nucl. Phys.* B **728** 121
[9] He X G, Lee C J, Li S F, and Tandean J 2014 *Phys. Rev.* D **89** 091901
[10] He X G, Lee C J, Li S F, and Tandean J 2014 *J. High Energy Phys.* JHEP08(2014)019
[11] Mohapatra R N *et al*, *Rept. Prog. Phys.* **70** 1757
[12] Olive K A *et al* 2014 *Chin. Phys.* C **38** 090001
[13] Casas J A and Ibarra A 2001 *Nucl. Phys.* B **618** 171
[14] Colangelo G, Nikolidakis E, and Smith C 2009 *Eur. Phys. J.* C **59** 75
[15] Mercolli L and Smith C 2009 *Nucl. Phys.* B **817** 1
[16] Capozzi F, Fogli G L, Lisi E, Marrone A, Montanino D, and Palazzo A, 2014 *Phys. Rev.* D **89** 093018

Journal of Physics: Conference Series **627** (2015) 012005 doi:10.1088/1742-6596/627/1/012005

Calculation for optimization of the experimental conditions for RBS analysis at the HUS 5SDH-2 tandem accelerator

Le Hong Khiem[1], Vi Ho Phong[2], Nguyen The Nghia[2]
[1]Institute of Physics, Vietnam Academy of Science and Technology, 10 Dao Tan, Ba Dinh, Hanoi, Vietnam
[2] Department of Nuclear Physics, Faculty of Physics, Hanoi University of Science, VNU, 334 Nguyen Trai, Thanh Xuan, Hanoi, Vietnam

E-mail: lhkhiem@iop.vast.ac.vn

Abstract. The dependences of the depth and mass resolutions of analysis using Rutherford Backscattering Spectrometry (RBS) on some experimental conditions (such as the beam energy, the target tilting angle, etc.) have been investigated. A computer program for simulating the RBS spectra and for calculating the depth and mass resolution under different experimental conditions was developed. The results of calculation were experimentally checked by using some reference samples. The good agreements between calculated and experimental values have been found. The optimum analysis conditions over a wide range of RBS applications based on our calculation can be chosen. This investigation was conducted by using the RBS system at HUS 5SDH-2 Tandem accelerator at the Hanoi University of Science.

1. Introduction

The HUS-5SDH-2 Tandem Pelletron accelerator system at Hanoi University of Science has been installed and started operating since 2012. It is mainly used for material characterization including RBS, which is very useful for probing the depth profile of near-surface regions of materials.

The purposes of this paper are: (i) to estimate the influence of experimental condition to some parameters by theoretical considerations, (ii) to measure the depth resolutions at various condition and compare them with the calculated values, (iii) to establish the optimum experimental condition. A computer program for calculation has been developed and used for estimation of several parameters such as the energy resolution, the accessible depth and the depth resolution that are under influence of experimental conditions.

All of experiments were carried out using RBS the Tandem 5SDH-2 Accelerator at Hanoi University of Science. In our experiment, RBS spectra collected with the thin gold films on glass substrate have been obtained with various experimental conditions such as energy of the beam and tilting angle of the samples. The calculations have been performed using our computer program at the same experimental conditions and the obtained results were compared with the experimental ones.

2. Theoretical calculations and the RBS simulation code

2.1. The mass resolution

The mass resolution δM_2 at mass M_2 is defined as the ability of the system which can distinguish a minimum value of mass difference. It is related to the separation of energy δE between signals in RBS spectra caused by different elements at certain depth as follow

$$\delta M_2 = \frac{\delta E_L}{E_0\left(\frac{dK}{dM_2}\right)}, \tag{1}$$

where K is the kinematic factor and the energy resolution δE_L at surface of the sample was obtained from analysis of RBS spectra in the low energy side.

2.2. *The accessible depth*

The accessible depth in RBS depends on the target compositions, tilting angle, energy of the incoming beam and mass. To satisfy the condition of accessible depth, the backscatted particles must emerges from the target with sufficient detectable energy E_1 . An arbitrary criterion has been proposed [8]

$$E_1 \geq \frac{1}{4} K E_0. \tag{2}$$

Here E_0 is the energy of incoming beam.

2.3. *The depth resolution*

The depth resolution at depth Nt is related to the minimum detectable energy difference δE_H of the scattered particles (which is referred as the total energy resolution) and is given by the following formula

$$\delta(Nt) = \delta E_H / \left(K\varepsilon_{in} \frac{1}{cos\alpha} + \varepsilon_{out} \frac{1}{cos\beta} \right), \tag{3}$$

where K is the kinematic factor, ε_{in} and ε_{out} are the stopping power of the incident and exit beam, α and β are the angles between target normal and incident and exit beam, respectively. The total energy resolution δE at certain depth is contributed from different sources including thedetector energy resolution, energy spread of incoming beam, energy straggling, geometry and multiple scattering effects. Detail expressions for those contributions are given in the appendix.

2.4. *The RBS simulation code*

A simple simulation computer code named RUT has been developed at our laboratory. The main part of this code is evaluation of energy E of particle having initial energy of E_0 after passing through a layer of material. This was calculated by solving the following first order ordinary differential equation by using adaptive step size Runge - Kutta algorithm [11]

$$\frac{dE}{dNt} = -\varepsilon(E), \tag{4}$$

with the initial value $E(Nt = 0) = E_0$. The stopping power ε can be calculated for a certain energy E using tabulated parameters from the electronic stopping power data by Andersen and Ziegler [2, 3]. The following parameters can be derived from the code: (i) The stopping powers at certain depth, (ii) the kinematic factor (iii) the accessible depth.

3. Experimental arrangement, data extraction

The RBS experiments were performed using the Pelletron Tandem accelerator model 5SDH-2 for thin gold film with thickness of 200 $\times 10^{15}$ cm^{-2} with the estimated standard error of 10% on glass substrate. The He beam after being accelerated by the Pelletron was focused, collimated and entered the RC43 scattering chamber equipped with a charged particles spectroscopy system. This system comprises a silicon surface barrier detector (energy resolution of 11 keV for 5.486 MeV alpha particle) and the conventional associated electronics. The experimental setup is shown in figure 1. The detector was placed at the distance of 130 mm from the target allowing for a detection solid angle of 3 mSr.

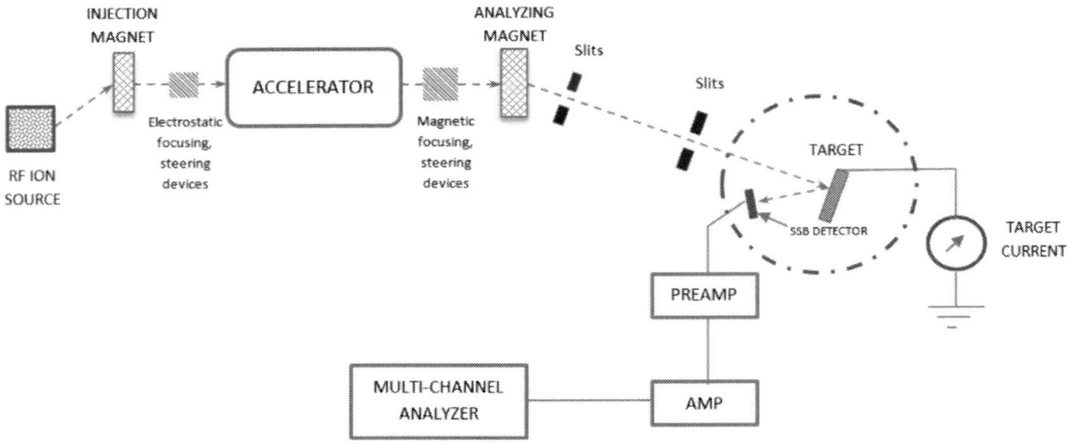

Figure 1. The schematic view of the accelerator and the RBS spectroscopy system equipped in the analytical beam line.

In our investigation only the depth resolution was compared with theoretical calculation. The depth resolution was determined from the spectra obtained with various energies of incident beam for the case of the target tilting angle of 50 degrees and various target tilting angle angles for the case of 2200 keV incident He ions.

The typical RBS spectrum which was used to extract the depth resolution is shown in figure 2. The backscatted signals from the thin Au film were numerically smoothed and their derivations were taken. The obtained spectra were fitted with the Gaussian function for low energy and high energy sides. The fitting interval starting from a point which corresponds 12% of full height and ending at a point which is corresponds 88% of full height was chosen [8]. The depth resolution at the depth Nt was determined by the following formula [6]

$$\delta(Nt) = \delta E_L \frac{Nt}{E_H - E_L}, \tag{5}$$

where E_L and E_H are the energy resolutions at low energy side and high energy side.

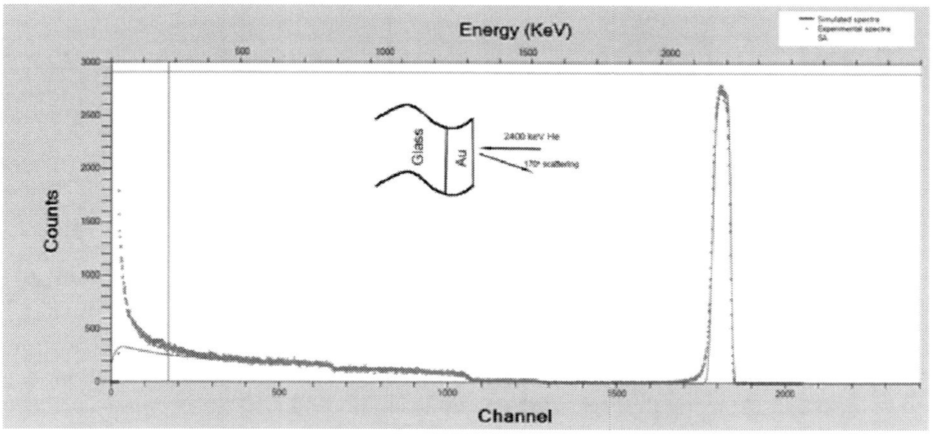

Figure 2. The typical RBS spectra for thin Au film on glass substrate. The high energy and low energy edges from backscatted signals of Au film are indicated.

4. Results and discussion

Figure 3 shows the simulated spectrum obtained by using our code for a sample which consists of some material layers together with a spectrum calculated by the well-known software RUMP [1] for the same sample. It is clearly seen from the figure that two spectra are in good agreement. Beside the simulation of the RBS spectrum, it is possible to use our simulation code to calculate several parameters including the kinematic factor, the accessible depth, the energy loss, the stopping range and the energy resolution at certain depth. From the results of these calculations, one can easily choose an optimal experimental setup in order to get the desired depth, best mass and depth resolutions.

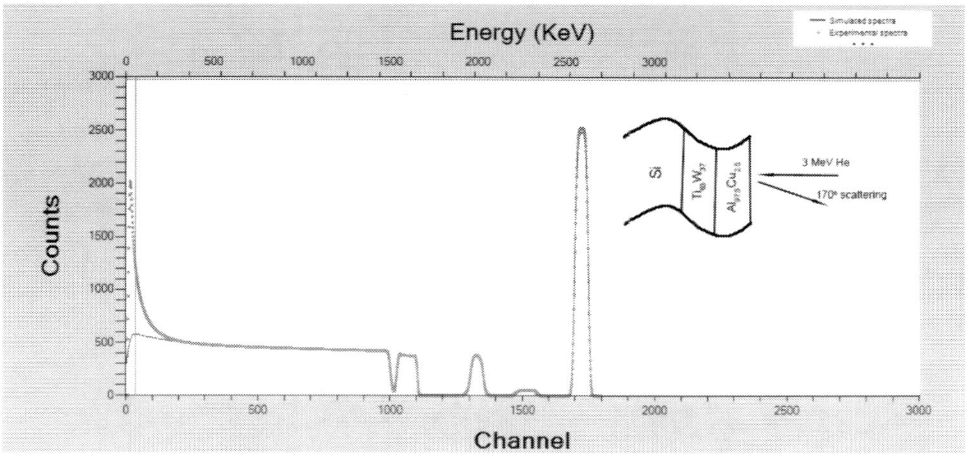

Figure 3. Simulated spectrum of $1.7x10^{18}$ cm^{-2} of an Al-Cu alloy on $0.55x10^{18}$ cm^{-2} of a Ti-W alloy on a silicon substrate generated by our simulation code (blue line) and RUMP code (red dots).

The theoretical values of the accessible depth calculated by our simulation code were compared with tabulated values taken from Ref. [8]. It is shown in table 1 that they are in good agreement for different beam types and beam energies.

Table 1.Accessible depth for several target material

Beam	Energy (keV)	Accessible depth (μm) (Calculated by our simulation code)				Accessible depth (μm) (Calculated by Chu [7])			
		Au	Ag	Ni	Al	Au	Ag	Ni	Al
^{4}He	1000	0.57	0.59	0.50	0.80	0.5	0.6	0.5	1
	2000	1.03	1.12	0.96	1.67	1	1.1	1.1	1.7
^{1}H	1000	2.38	2.81	2.44	5.40	2.4	2.7	2.4	4.5
	2000	6.47	8.12	6.91	15.94	6.4	7.7	6.9	15

For determination of mass resolution, the energy resolution at different incident energies was measured by backscattering experiment using a thin Au target. The mass resolution was then determined by equation (1) using the measured energy resolution. Figure 4 shows the dependences of obtained mass resolution as well as the energy resolution on the energy of ^{4}He beam. It is seen form figure 4 that the energy resolution does not change when the energy of the incident beam increases and the corresponding obtained mass resolution decreases with the increase of energy of the incident beam. By using this procedure, one can determine the optimized energy of the incident beam in order to get the best mass resolution.

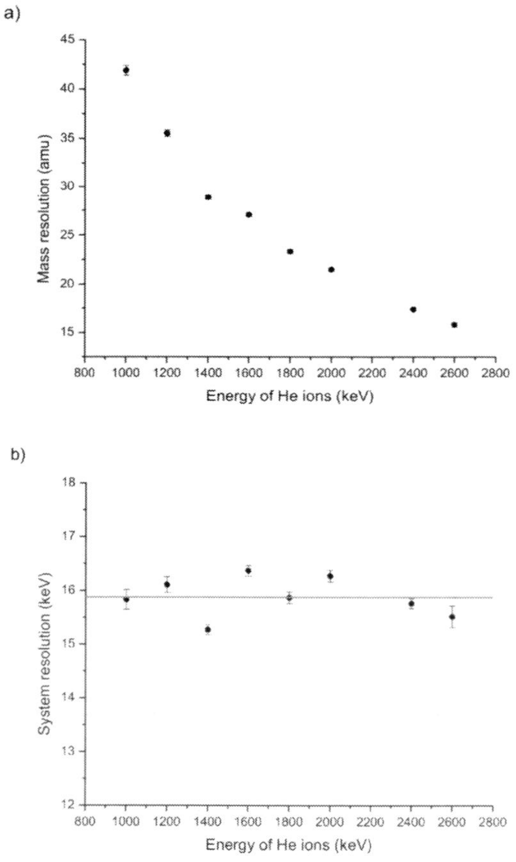

Figure 4. Experimental mass resolution (a) and corresponding energy resolution (b) as a function of energy of He ions.

The depth resolution for a specific energy of the beam and target tilting angle can be calculated by equation (3). For this purpose our simulation code provides an estimation of the stopping power and the total energy resolution that takes the energy resolution of the beam, the energy resolution of the detector and the energy straggling effect into account. The calculated values of the depth resolution were verified by backscattering experiment using a thin Au film at varying beam energies and target tilting angles. The uncertainty of the experimental values were less than 10.2%. Figure 5 shows the measured and calculated depth resolution at depth 200 $x10^{15}$ atom/cm^2 for Au film as a function of the energy of the He beam and target tilting angles. For different beam energies, the measured depth resolutions are in agreement with the calculated values within 8.3%. The measured values for different tilting angles also fairly agreed with calculated value at the small angles up to 50 degrees. However, there are some noticeable discrepancies between measured and calculated values for large tilting angles. These discrepancies might be explained by the fact thatthe effects of large angle multiple scattering and lateral spreads were not included in the calculation. By using these obtained results we were able to choose the optimal beam energy as 1200 keV and the optimal target tilting angle of 50 degree.

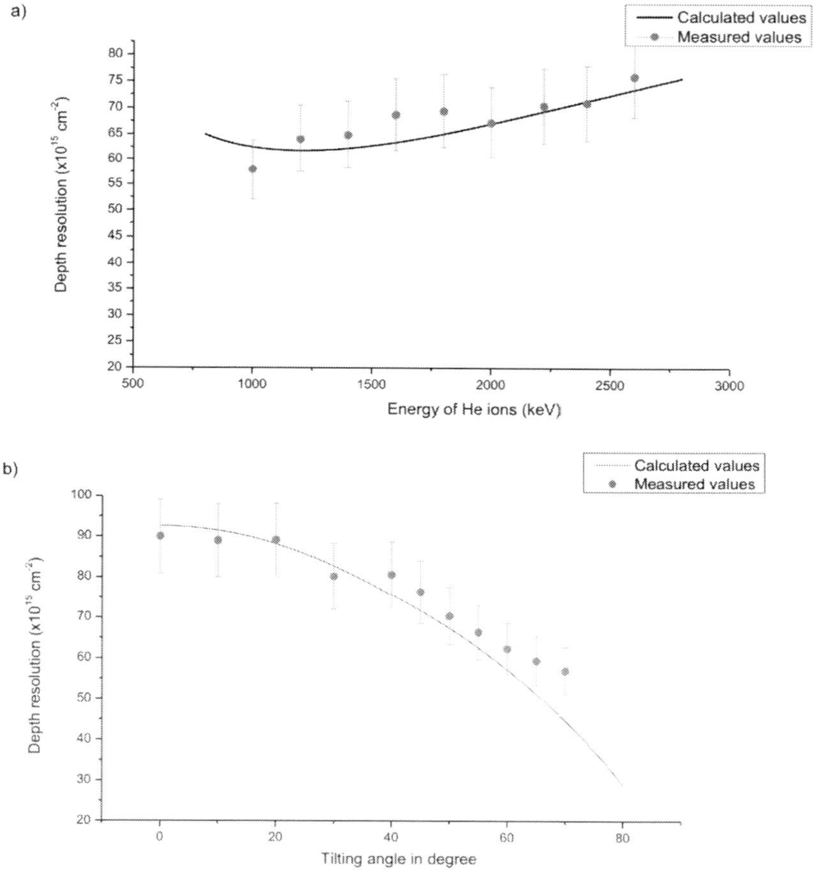

Figure 5. Depth resolution at depth of $200x10^{15}$ cm^{-2} for Au sample as a function of the energy of the incident He ions in the case of target tilting angle of 50 degrees (a) and as a function of target tilting angles in the case of 2200 keV He ions (b).

The above optimization procedures were also successfully applied for a W implanted on Kapton sample having a depletion region of Oxygen caused by W layer. Two RBS spectra are presented in figure 6. The spectra on the left and right panels were obtained by using non-optimized and optimized experimental conditions respectively. It is seen from this figure that the mass resolution of the RBS spectrum with optimized experimental condition on the right side is better than that of non-optimized experimental condition.

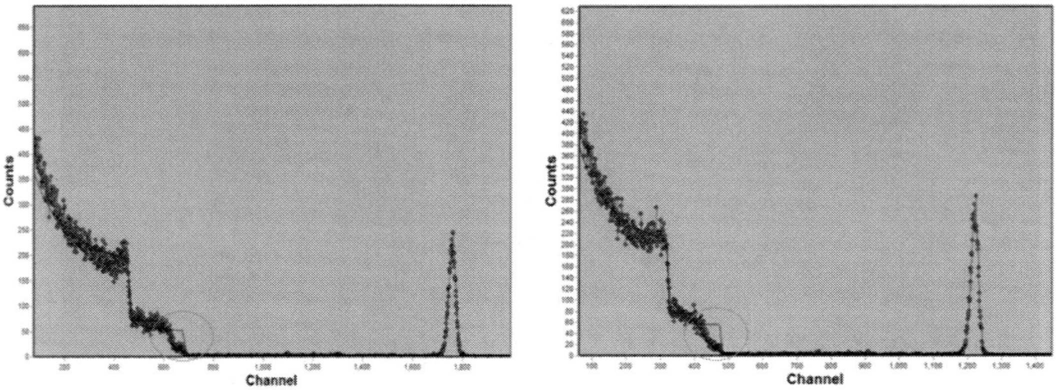

Figure 6. The RBS spectra for the W implanted on Kapton sample using a non-optimized experimental condition (left) and an optimized experimental condition (right). The green marked region indicates effect of depletion of Oxygen from Kapton after implantation by a high dose of W

5. Conclusion

The dependences of the depth and mass resolutions of analysis using Rutherford Backscattering Spectrometry (RBS) on some experimental conditions (such as the beam energy, the target tilting angle, etc.) have been investigated. A computer program for simulating the RBS spectra and for calculating the depth and mass resolution under different experimental conditions was developed. The results of calculation were experimentally checked by using some reference samples. The good agreements between calculated and experimental values have been found. The optimum analysis conditions over a wide range of RBS applications based on our calculation can be chosen. This investigation was conducted by using the RBS system at HUS 5SDH-2 Tandem accelerator at the Hanoi University of Science. This work is supported by Hanoi University of Science.

Appendix

The energy resolution at surface of the sample in equation (1) is calculated as follows [12]

$$\delta E_H = [(\delta E_{BE})^2 + (\delta E_{BS})^2 + (\delta E_{BW})^2 + (\delta E_{SR})^2 + (\delta E_{AA})^2 + (\delta E_D)^2]^{1/2}, \quad \text{(A.1)}$$

where δE_{BE} is the energy spread of the incident ion beam; δE_{BS} is the angular spread of the incident ion beam; δE_{BW} is the energy spread due to beam profile on the target; δE_{SR} is due to the surface roughness of the sample; δE_{AA} is due to the detector acceptance angle and δE_D is the energy resolution of the detector. In our calculation only δE_D, δE_{BE} and δE_{SR} were considered. Therefore the energy resolution at a certain depth is given as

$$\delta E_L = \left(\delta E_H{}^2 + \delta E_{ES}{}^2 + \delta E_{MS}{}^2 + \delta E_{LS}{}^2\right)^{1/2}. \quad \text{(A.2)}$$

The following formula was used to calculate the energy straggling δE_{ES} of the incoming beam in solid

$$\delta E_{ES} = 2\left[2\ln 2\right]^{1/2}\left(K^2\Omega_{in}{}^2 + \Omega_{out}{}^2\right)^{1/2}, \tag{A.3}$$

where K is the kinematic factor. $\Omega_{in/out}$ are calculated for incoming and outgoing path of the beamby using the well-known Bohr formula [7]

$$\Omega_B = 4\pi\left(Z_1 e^2\right)^2 Z_2 Nt, \tag{A.4}$$

where Nt is the path length, Z_1, Z_2 the atomic number the projectile and the target, respectively.

References

[1] Lawrence R Doolittle 1985 *Nucl. Instr. and Meth.* B. **9** 344-351
[2] Ziegler J F 1977 *Helium - Stopping Powers and Ranges in All Elements, vol. 4 of The Stopping and Ranges of Ions in Matter* (New York: Pergamon Press)
[3] Ziegler J F and Andersen H H 1977 *Hydrogen - Stopping Powers and Ranges in All Elements vol. 3 of The Stopping and Ranges of Ions in Matter* (New York: Pergamon Press)
[4] Williams J S and Moller W 1978 *Nucl. Instr. and Meth.* **157** 213-221
[5] Williams J S 1978 *Nucl. Instr. and Meth.* **149** 207
[6] Kim M S *et al* 1996 *Nucl. Instr. and Meth.* B **108** 136-146
[7] Bohr N and Dan Vidensk Selsk K 1948 *Mat.-Fys. Medd* **18** 8
[8] Chu W K, Mayer J W and Nicolet M A 1978 *Backscatering Spectrometry* (New York:Academic Press)
[9] Yoshiaki Kido and Yukio Oso 1985 *Nucl. Instr. and Meth.* B **9** 291-300
[10] Yongqiang Wang and Michael Nastasi 2009 *Hanbook of Mordern Ion Beam Materials Analysis Second Edition* (Pennsylvania: Material Research Society)
[11] Press W H, Flannery B P, Teukolsky S A and Vetterling W T 1988 *Numerical Recipes* (New York:Cambridge University Press)
[12] O'connor D J and Tan Chunyu 1989 *Nucl. Instr. and Meth.* B **36** 178-188

Journal of Physics: Conference Series **627** (2015) 012006 doi:10.1088/1742-6596/627/1/012006

Reentrance phenomenon of superfluid pairing in hot rotating nuclei

N Quang Hung[1], **N Dinh Dang**[2], **B K Agrawal**[3], **V M Datar**[4], **A Mitra**[4], **and D R Chakrabarty**[4]

[1] School of engineering, Tan Tao University, Tan Tao University Avenue, Tan Duc E.City, Duc Hoa, Long An Province, Vietnam

[2] Theoretical Nuclear Physics Laboratory, RIKEN Nishina Center for Accelerator-Based Science, 2-1 Hirosawa, Wako City, 351-0198 Saitama, Japan and Institute for Nuclear Science and Technique, Hanoi, Vietnam

[3] Theory Division, Saha Institute of Nuclear Physics, 1/AF, Bidhan Nagar, Kolkata 700 064, India

[4] Nuclear Physics Division, Bhabha Atomic Research Center, Mumbai 40085, India

E-mail: `hung.nguyen@ttu.edu.vn (N.Q.H)`

Abstract. Reentrance phenomenon of superfluid pairing in highly rotating excited (hot) nuclei is studied within the Bardeen-Cooper-Schrieffer (BCS)-based approach (the FTBCS1), taking into account the effect of quasiparticle number fluctuations (QNF) on the pairing field at finite temperature and angular momentum within the pairing model plus noncollective rotation along the symmetry axis. The numerical calculations are carried out for the pairing gaps and level densities in ^{104}Pd, of which an anomalous enhancement of level density at low excitation energy E^* and high angular momentum J has been experimentally observed. The results obtained show that the QNF within the FTBCS1 theory smooth out the superfluid-normal phase transition and lead to the appearance of pairing reentrance phenomenon in hot rotating nuclei. This feature can be clearly seen in the behavior of pairing gap obtained at low E^* and high J. The good agreement between the level densities obtained within the FTBCS1 and those extracted from the experiment indicates that the observed enhancement in the level densities of ^{104}Pd nucleus is a manifestation of the pairing reentrance phenomenon.

1. Introduction

When a nucleus rotates (total angular momentum J and/or rotational frequency ω are not zero), the nucleon (proton and neutron) pairs located around the Fermi surface will scatter to the empty levels nearby and lead to the decreasing of pairing correlation. When the J or ω is sufficiently high, i.e., equal to the critical value J_c or ω_c, the scattered nucleons block completely the single-particle levels around the Fermi surface. Consequently, pairing correlation disappears. This phenomenon is called the Mottelson-Valatin effect [1], which is qualitatively similar as the collapsing of pairing correlation at the critical temperature T_c, where the superfluid-normal (SN) phase transition occurs. However, when J is slightly higher than J_c (or $\omega \geq \omega_c$), the increase of temperature T will relax the particles scattered around the Fermi surface and causes some levels become partially unoccupied, and therefore available for scattered pairs. As a result, when T increases up to a critical value T_1, the pairing correlation reappears. As T goes higher, e.g., at $T_2 > T_1$, the newly created pairs will again break and therefore the pairing correlation

Journal of Physics: Conference Series **627** (2015) 012006 doi:10.1088/1742-6596/627/1/012006

is eventually broken down. This phenomenon, which was first predicted by Kammuri [2] and Moretto [3], is called thermally assisted pairing correlation or anomalous pairing, and later as pairing reentrance by Balian, Flocard, and Veneroni [4]. It is similar to the Meissner effect in the superconducting metal in the presence of an external magnetic field, where the magnetic field plays the role as that of nuclear rotation [5].

It has been shown in the 1960s that the SN phase transition is an artifact of the BCS method because it neglects the thermal fluctuations in finite systems. The latter have been shown to be large so that they smooth out the SN phase transition in finite nuclei [6]. As the result, the pairing gap does not collapse at T_c , but monotonically decreases with increasing T and remains finite at $T \geq T_c$. The calculations within an exactly solvable pairing Hamiltonian for clusters and nuclei [7] and the cranked deformed shell model [8] for a single j-shell have predicted the pairing reentrance effect but in the way that pairing gap, which is zero at $J > J_c$ and $T = 0$, reappears at a certain T and does not vanish as predicted by the conventional FTBCS theory. The recently developed FTBCS1 theory that includes the effect due to quasiparticle-number fluctuations in the pairing field and angular momentum z projection at $T \neq 0$ has also predicted a similar behavior of pairing reentrance effect in some realistic nuclei [9, 10, 11]. In addition, the shell-model Monte Carlo calculations for heated rotating ^{72}Ge nucleus have suggested for the first time that the pairing reentrance effect can be seen not only in the behavior of pairing gap as functions of T and J (or ω) but also in the nuclear level density, an experimentally observed quantity, in a form of a local maximum at low T (or excitation energy E^*) and high J (or ω) [12]. Recently, by fitting the proton spectra of ^{12}C + ^{93}Nb $\to$ ^{105}Ag* $\to$ ^{104}Pd* + p reaction at the incident energy of 40 - 45 MeV, an enhancement of level density of ^{104}Pd at low E^* and high J (approximately greater than 17 $\hbar$) has been reported [13]. This enhancement is qualitatively similar to that predicted by the shell-model Monte Carlo simulation for ^{72}Ge and therefore might come from the pairing reentrance. The goal of the present work is to answer the question whether the enhancement observed in the extracted level density of ^{104}Pd is the first evidence of pairing reentrance phenomenon in atomic nuclei. For that purpose, we employ the FTBCS1 theory at finite temperature and angular momentum, in which the thermal fluctuations are included.

2. FTBCS1 theory at finite temperature and finite angular momentum

The FTBCS1 equations at finite temperature and angular momentum are derived based on the variational method to minimize the expectation value of the pairing Hamiltonian

$$H = \sum_k \epsilon_k (a^\dagger_{+k} a_{+k} + a^\dagger_{-k} a_{-k}) - G \sum_{kk'} a^\dagger_k a^\dagger_{-k} a_{-k'} a_{k'} - \lambda \hat{N} - \omega \hat{M} \ , \tag{1}$$

in the grand-canonical ensemble [3, 10]. Here, the Hamiltonian (1) describes a system rotating about the symmetry axis, which is chosen to coincide with its z component. The particle-number operator $\hat{N}$ and the z projection $\hat{M}$ of the total angular momentum $\hat{J}$ (which coincides with $\hat{M}$ for spherical nuclei) are defined as

$$\hat{N} = \sum_k (a^\dagger_{+k} a_{+k} + a^\dagger_{-k} a_{-k}) \ , \quad \hat{M} = \sum_k m_k (a^\dagger_{+k} a_{+k} - a^\dagger_{-k} a_{-k}) \ , \tag{2}$$

where $a^\dagger_{\pm k} (a_{\pm k})$ are the creation (annihilation) operators of a particle in the k-th deformed state, whereas ϵ_k, λ, and ω are respectively the single-particle energies, chemical potential, and rotational frequency. The FTBCS1 equation for the pairing gap has the final form as [10, 11]

$$\Delta_k = \Delta + \delta \Delta_k \ , \tag{3}$$

where

$$\Delta = G\sum_{k'} u_{k'}v_{k'}(1-n_{k'}^{+}-n_{k'}^{-}) , \quad \delta\Delta_k = G\frac{\delta\mathcal{N}_k^2}{1-n_k^{+}-n_k^{-}}u_kv_k , \tag{4}$$

$$\begin{aligned} u_k^2 &= \frac{1}{2}\left(1+\frac{\epsilon_k - Gv_k^2-\lambda}{E_k}\right) , \quad v_k^2 = \frac{1}{2}\left(1-\frac{\epsilon_k - Gv_k^2-\lambda}{E_k}\right) , \\ E_k &= \sqrt{(\epsilon_k - Gv_k^2-\lambda)^2+\Delta_k^2} , \quad n_k^{\pm} = \frac{1}{1+e^{\beta(E_k\mp\omega m_k)}} , \quad \beta = 1/T , \end{aligned} \tag{5}$$

with $\delta\mathcal{N}_k^2$ being the quasi-particle-number fluctuations (QNF)

$$\delta\mathcal{N}_k^2 = (\delta\mathcal{N}_k^{+})^2 + (\delta\mathcal{N}_k^{-})^2 = n_k^{+}(1-n_k^{+}) + n_k^{-}(1-n_k^{-}) . \tag{6}$$

The equations for the particle number and total angular momentum are given as

$$N = 2\sum_k\left[v_k^2(1-n_k^{+}-n_k^{-})+\frac{1}{2}(n_k^{+}+n_k^{-})\right], \quad M = \sum_k m_k(n_k^{+}-n_k^{-}) . \tag{7}$$

In Eq. (4), if $\delta\mathcal{N}_k^2$ is set to be zero, i.e., no QNF, one recovers the conventional FTBCS equations at finite T and M. The total level density of a system with N neutrons and Z protons and energy $\mathcal{E}$ is obtained by using the inverse Laplace transformation of the grand partition function [3]. It reads

$$\rho(\mathcal{E}, M) = \frac{e^{(S_N+S_Z)}}{(2\pi)^2\sqrt{D}} , \tag{8}$$

$$D = \begin{vmatrix} \frac{\partial^2\Omega}{\partial\alpha_N^2} & \frac{\partial^2\Omega}{\partial\alpha_N\partial\alpha_Z} & \frac{\partial^2\Omega}{\partial\alpha_N\partial\mu} & \frac{\partial^2\Omega}{\partial\alpha_N\partial\beta} \\ \frac{\partial^2\Omega}{\partial\alpha_Z\partial\alpha_N} & \frac{\partial^2\Omega}{\partial\alpha_Z^2} & \frac{\partial^2\Omega}{\partial\alpha_Z\partial\mu} & \frac{\partial^2\Omega}{\partial\alpha_Z\partial\beta} \\ \frac{\partial^2\Omega}{\partial\mu\partial\alpha_N} & \frac{\partial^2\Omega}{\partial\mu\partial\alpha_Z} & \frac{\partial^2\Omega}{\partial\mu^2} & \frac{\partial^2\Omega}{\partial\mu\partial\beta} \\ \frac{\partial^2\Omega}{\partial\beta\partial\alpha_N} & \frac{\partial^2\Omega}{\partial\beta\partial\alpha_Z} & \frac{\partial^2\Omega}{\partial\beta\partial\mu} & \frac{\partial^2\Omega}{\partial\beta^2} \end{vmatrix} . \tag{9}$$

where the grand-partition function Ω, total energy $\mathcal{E}$, and entropy S are given as

$$\Omega = \Omega_N + \Omega_Z = S_N + S_Z + \alpha_N N + \alpha_Z Z + \mu M - \beta\mathcal{E} , \quad \mathcal{E} = < H > = \frac{\partial\Omega}{\partial\beta} , \tag{10}$$

$$S = -\sum_k [n_k^{+}\ln n_k^{+} + (1-n_k^{+})\ln(1-n_k^{+}) + n_k^{-}\ln n_k^{-} + (1-n_k^{-})\ln(1-n_k^{-})] , \tag{11}$$

where $\alpha = \beta\lambda$ and $\mu = \beta\omega$. The total level density $\rho(\mathcal{E})$ is indeed calculated as the sum of J-dependent level densities, namely $\rho(\mathcal{E}) = \sum_J(2J+1)\rho(\mathcal{E}, J)$ [14], where $\rho(\mathcal{E}, J)$ is obtained by differentiating $\rho(\mathcal{E}, M)$ [15]

$$\rho(\mathcal{E}, J) = \rho(\mathcal{E}, M = J) - \rho(\mathcal{E}, M = J+1) . \tag{12}$$

3. Results

The numerical calculations are carried out for ^{104}Pd nucleus, for which the single-particle spectra are taken from the axially deformed Woods-Saxon potentials with the Blomqvist-Wahlborn parametrization at a fixed value of quadrupole deformation parameter $\beta_2 = 0.276$ [16]. The pairing constant G is adjusted so that the pairing gaps obtained within the FTBCS (FTBCS1) at $T = 0$ for proton and neutron fit the empirical odd-even mass differences, namely

$\Delta_N = 1.26$ MeV and $\Delta_Z = 1.5$ MeV [17]. Figures 1 (a)-(b) show that the level weighted pairing gaps $\bar{\Delta} \equiv \sum_k \Delta_k/\Omega$ (with Ω being the sum of all single-particle levels) obtained within the conventional FTBCS (thin lines) at all J decrease with increasing E^* and collapses at some critical values E^*_c. Here E^* is calculated as $E^* = \mathcal{E}(T, M) - \mathcal{E}(0, M)$. There is no significant enhancement of the level densities obtained within the FTBCS [See e.g., dotted lines in Figs. 1 (c)-(f)] as those in the experimental data at $J \geq 20\hbar$. As a result, the FTBCS does not show the pairing reentrance effect at all J.

For the FTBCS1, due to the inclusion of the QNF in the gap equation (4), the FTBCS1 gaps decrease monotonically with increasing E^* and do not collapse at $E^* = E^*_c$ as in the case of the FTBCS. Instead the pairing gaps in the FTBCS1 remain finite even at $E^* > 15$ MeV. Within the FTBCS1, the pairing reentrance is seen very clearly at $J = 20\hbar$ for neutrons and at $J = 30\hbar$ for protons [See e.g., dashed lines in Figs. 1 (c)-(f)]. Consequently, there appear local enhancements in the FTBCS1 level densities at around $2 < E^* < 5$ MeV at these two values of J. The FTBCS1 level densities agree fairly well with the experimental data at all J values considered in present work.

It is worth noticing that the FTBCS1 employed in this work includes only the monopole pairing, whereas the higher multipolarities, which are responsible for the collective motion in finite nuclei and might lead to the increase of the level densities, are neglected in the residual interaction. We therefore do not expect an excellent quantitative agreement between the theoretical predictions and the experimental data. However, the qualitative physics in the present work is not affected by this neglect, that is, the enhancement observed in the experimentally extracted level density at low E^* and high J in ^{104}Pd nucleus is indeed an evidence of pairing reentrance.

4. Conclusions

The present work employs the FTBCS1 theory at finite temperature and angular momentum to study the pairing phenomenon and level density in ^{104}Pd, of which an enhancement of level density at low excitation energy and high angular momentum has been experimentally observed. The quantitative agreement between experiment and theory suggests that this enhancement is indeed the first experimental evidence of the reentrance of superfluid pairing in a finite nucleus. The numerical calculations were carried out using the Integrated Cluster of Clusters (RICC) system at RIKEN. This work is supported by the National Foundation for Science and Technology Development (NAFOSTED) of Vietnam through Grant No.103.04-2013.08.

References

[1] B.R. Mottelson B R and Valatin J G 1960, *Phys. Rev. Lett.* **5** 511.
[2] Kammuri T 1964, *Pro. Theor. Phys.* **31** 595.
[3] Morreto L G 1972, *Nucl. Phys.* **A185** 145.
[4] Balian R, Flocard H and Veneroni M 1999, *Phys. Rep.* **317** 251.
[5] Levy F, Sheikin I, Grenier B, and Huxley A D 2005, *Science* **309** 1343.
[6] Mang H J, Rasmussen O, and Rho M 1996, *Phys. Rev.* **141** 941; Moretto L G 1972, *Phys. Lett.* **B40** 1; Goodman A L 1984, *Phys. Rev.* **C29** 1887; Dang N D and Zelevinsky V 2001, *Phys. Rev.* **C64** 064319; Dang N D 2007, *Nucl. Phys.* **A784** 147; Zelevinsky V, Brown B A, Fazier N, and Horoi M 1996, *Phys. Rep.* **276** 85.
[7] Frauendorf S, Kuzmenko N K, Mikhajlov V M, and Sheikh J A 2003, *Phys. Rev.* **B68** 024518.
[8] Sheikh J A, Palit R, and Frauendorf S 2005, *Phys. Rev.* **C72** 041301(R).
[9] Dang N D and Hung N Q 2008, *Phys. Rev.* **C77** 064315.
[10] Hung N Q and Dang N D 2008, *Phys. Rev.* **C78** 064315.
[11] Hung N Q and Dang N D 2011, *Phys. Rev.* **C84** 054324.
[12] Dean D J, Langanke K, Nam H A, and Nazarewicz W 2010, *Phys. Rev. Lett.* **105** 212504.
[13] Mitra A *et al.* 2009, *J. Phys. G* **36** 095103; Mitra A *et al.* 2010, *EPJ Web of Converences* **2** 04004; Datar V, Mitra A, and Chakrabarty D R, *unpublished*.
[14] Gilbert A and Cameron A G W 1965, *J. Phys.* **43** 1446.

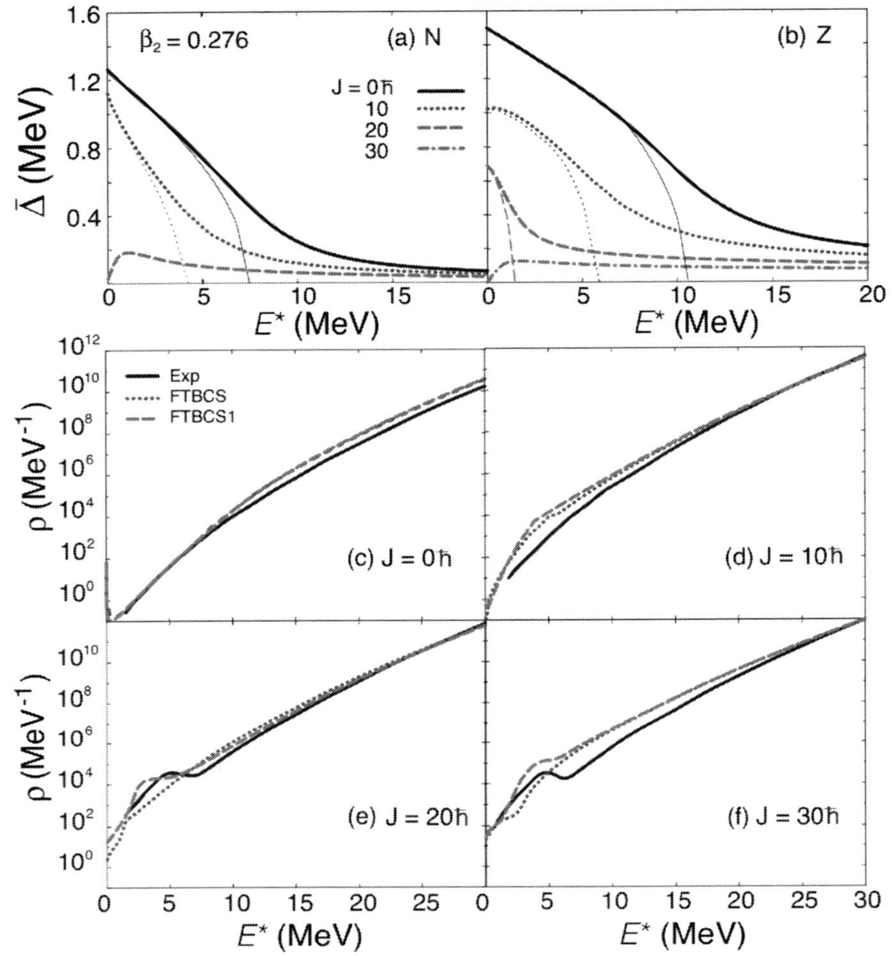

Figure 1. . Level-weighted pairing gaps $\bar{\Delta}$ for neutron (N) (a) and protons (Z) (b) and total level densities (c) - (f) as function of excitation energy E^* obtained within the FTBCS and FTBCS1 at the quadrupole deformation parameter $\beta_2 = 0.276$. The thin and thick lines in (a) and (b) denote the FTBCS and FTBCS1 results, respectively, whereas the dotted and dashed lines in (c) - (f) respectively stand for the FTBCS and FTBCS1 total level densities. The solid lines in (c) - (f) are the experimentally extracted data.

[15] Ericson T 1960, *Adv. Phys.* **9** 425.
[16] Cwiok S *et al.* 1987, *Com. Phys. Comm.* **46** 379.
[17] Ring P and Schuck P 2004, *The Nuclear Many-Body Problem*, Springer, Heidelberg.

Journal of Physics: Conference Series **627** (2015) 012007 doi:10.1088/1742-6596/627/1/012007

The metric of extended Einstein equation and Schwarzschild solution in six dimensions

Phan Hong Lien[1], **Tran Huu Phat**[2] **and Ha Van Chung**[3]

[1] Le Qui Don Academy of Technology, 236 Hoang Quoc Viet, Hanoi, Vietnam
[2] Vietnam Atomic Energy Institute, 59 Ly Thuong Kiet, Hanoi, Vietnam
[3] Hanoi National University of Education, 136 Xuan thuy, Hanoi, Vietnam

E-mail: phlien.mta@gmail.com, tranhuuphat41@gmail.com

Abstract. The metric of a warped space-time with two extra dimensions is established by means of the Einstein equations in six dimensions and the compactification of two extra dimensions on a square. It is shown that at every among two fixed points our manifold reduces to the five-dimensional Randall - Sundrum space-time and the hierarchy problem could be solved. The Schwarzschild solutions have been extended in six dimensions. We also proposed an equation for the mass operator generated from extra dimensional coordinates.

1. Introduction

Besides the extended models of Standard Model (SM), the extra dimensions have been considered as the ways to investigate both well-studied and new problems of particle physics and Cosmology, in the new physics beyond the SM to unify gravity with gauge interactions. Extra dimensional theories can lead to really significant physical consequences.

We started with the following important observations

(i) Solving the mass hierarchy problem by means of the wrap factor attaching to the four - dimensional metric is very successful [1,2];

(ii) The proton stability can be explained within the framework of a theory with two extra dimensions [3,4] that ensures a lifetime of proton longer than the current experimental bounds [4], even in the presence of baryon violation at TeV scale, and forces the neutrino masses to be of the Dirac type [5]. It also explains the origin of electroweak symmetry breaking [6], the number of fermion generation [7] and the breaking of grand unified gauge groups [8-10].

(iii) The existence of chiral fermions [3] in the four - dimensional effective theory has good prospect for setting up the six-dimensional standard model within a warped space-time.

We assume that the geometry of our theory is a 6D Riemann manifold, whose metric is expressed as

$$ds^2 = G_{AB}dx^A dx^B = e^{-2f(y_1,y_2)}g_{\mu\nu}dx^\mu dx^\nu - dy_1^2 - dy_2^2, \tag{1}$$

where $g_{\mu\nu} = diag(1,-1,-1,-1)$ is the 4- dimensional tensor metric. The coordinates are $x^A = x^\mu, y_1, y_2$ with $\mu,\nu = 0,1,2,3$ labelling the brane directions, $0 \leq y_1, y_2 \leq L$.

The general 6- dimensional action takes the form

$$\begin{aligned} S =& S_{grav} + S_{hid} + S_{vis} \\ =& \int d^4x \int dy_1 dy_2 \sqrt{-G}\Big[(-\Lambda + 2M^4 R) \\ &- \sqrt{-g_{vis}} V_{vis} \delta(y_1 - L)\delta(y_2 - L) - \sqrt{-g_{hid}} V_{hid} \delta(y_1)\delta(y_2)\Big]. \end{aligned} \tag{2}$$

The 6- dimensional Einstein's Equation reads

$$\begin{aligned} \sqrt{-G}\left(R_{AB} - \frac{1}{2} G_{AB} R\right) =& \frac{1}{4M^4}\Big[\Lambda\sqrt{-G} G_{AB} \\ &+ V_{vis}\sqrt{-g_{vis}} g^{vis}_{\mu\nu} \delta^\mu_A \delta^\nu_B \delta(y_1 - r_c)\delta(y_2) \\ &+ V_{hid}\sqrt{-g_{hid}} g^{hid}_{\mu\nu} \delta^\mu_A \delta^\nu_B \delta(y_1)\delta(y_2)\Big], \end{aligned} \tag{3}$$

where the $(3+1)$ - dimensional space - time is connected with the $(4+2)$ - dimensional by

$$\begin{aligned} g^{vis}_{\mu\nu}(x) =& G_{AB}\left(x, y_1 = L, y_2 = L\right), \\ g^{hid}_{\mu\nu}(x) =& G_{AB}\left(x, y_1 = 0, y_2 = 0\right). \end{aligned} \tag{4}$$

Our aim was to determine the analytical expression for $f(y_1, y_2)$ so that above metric is a solution of Einstein's equation extended in six dimensions. We also obtained the physical consequences when the 6D metric was considered by the 4D effective theory. Similarly in the case of RS space-time, the hierarchy problem was solved due to the Planck mass. We also consider the metric in the central symmetric system of 6D coordinates. In addition, the equation for mass operator in extra dimensions was proposed.

This paper is organized as follows. In Section 2, the Einstein equations extended in 6D space-time are solved in the cylindrical coordinate system. Section 3 is devoted to investigate the hierarchy problems. The Schwarzchild solutions in 6D are extended in Section 4. In Section 5, we present an equation for mass operator inclusive extra dimensional coordinates. The discussions and conclusions are given in Section 6.

2. The six-dimensional solution of Einstein's equation in the cylindrical symmetry

We impose two external variables in the cylindrical coordinates system

$$y_1 = \rho\cos\theta; y_2 = \rho\sin\theta, \tag{5}$$

where $0 \le \rho \le r_c, -\pi \le \theta \le \pi, f(y_1, y_2) = f(\rho, \theta)$.

This leads to the following metric

$$ds^2 = e^{-2f(\rho,\theta)} g_{\mu\nu} dx^\mu dx^\nu - d\rho^2 - \rho^2 d\theta^2. \tag{6}$$

We determine the bulk metric which is the solution of 6 - dim Einstein's equation

$$\begin{aligned} &\sqrt{-G}\left(R_{AB} - \frac{1}{2} G_{AB} R\right) = \frac{1}{4M^4}\Big[\Lambda\sqrt{-G} G_{AB} \\ &+ V_{vis}\sqrt{-g_{vis}} g^{vis}_{\mu\nu} \delta^\mu_A \delta^\nu_B \delta(\rho - r_c)\delta(\theta - \pi) + V_{hid}\sqrt{-g_{hid}} g^{hid}_{\mu\nu} \delta^\mu_A \delta^\nu_B \delta(\rho)\delta(\theta)\Big], \end{aligned} \tag{7}$$

where

$$\begin{aligned} g^{vis}_{\mu\nu}(x^\mu) =& G_{AB}\left(x^\mu, \rho = r_c, \theta = \pi\right), \\ g^{hid}_{\mu\nu}(x^\mu) =& G_{AB}\left(x^\mu, \rho = 0, \theta = 0\right), \end{aligned} \tag{8}$$

Journal of Physics: Conference Series **627** (2015) 012007 doi:10.1088/1742-6596/627/1/012007

and

$$G = detG_{AB}; \sqrt{-G} = r_c e^{-4f(\rho,\theta)},$$
$$R = 4\left[2f"_\rho(\rho,\theta) + \frac{2}{r_c^2}f"_\theta(\rho,\theta) - 5f_\rho'^2(\rho,\theta) - \frac{5}{r_c^2}f_\theta'^2(\rho,\theta)\right].$$

By separating the indices $\mu\nu$, $\theta\theta$, and $\rho\rho$, we derive the system of equations from the Einstein equation (7)

$$4f"_\rho(\rho,\theta) + 10f_\rho'^2(\rho,\theta) + \frac{6}{r_c^2}f_\theta'^2(\rho,\theta) = -\frac{\Lambda}{4M^4}, \quad (9)$$

$$-\frac{4}{r_c^2}f"_\theta(\rho,\theta) + \frac{10}{r_c^2}f_\theta'^2(\rho,\theta) + 6f_\rho'^2(\rho,\theta) = -\frac{\Lambda}{4M^4}, \quad (10)$$

$$\begin{aligned} &3f"_\rho(\rho,\theta) + \frac{3}{r_c^2}f"_\theta(\rho,\theta) - 6f_\rho'^2(\rho,\theta) - \frac{6}{r_c^2}f_\theta'^2(\rho,\theta), \\ &= -\frac{\Lambda}{4M^4} + \frac{V_{vis}}{4M^4 r_c}\delta(\rho - r_c)\delta(\theta - \pi) + \frac{V_{hid}}{4M^4 r_c}\delta(\rho)\delta(\theta). \end{aligned} \quad (11)$$

It is clear that when ρ coordinate is absent, i.e. the sixth dimension disappears, Eq. (9) - (11) reduce to the Einstein's Eq. in the RS space - time.

Here the single scale k is related to cosmological constant Λ by

$$\Lambda = -24k^2M^4, \quad (12)$$

and

$$V_{vis} = -24M^4k; \quad V_{hid} = 24M^4k. \quad (13)$$

The solution for the bulk metric reads

$$ds^2 = e^{-2k\rho}g_{\mu\nu}dx^\mu dx^\nu - d\rho^2 - \rho^2 d\theta^2. \quad (14)$$

Generally, the solution - metric of Einstein's Equation in 6 dimensional space - time takes the form

$$ds^2 = e^{-2k|z|}g_{\mu\nu}dx^\mu dx^\nu - dy_1^2 - dy_2^2, \quad (15)$$

where $z = y_1 + iy_2$.

It is clear that (15) will reduce to the RS metric in the following cases:

(i) Compactification on a 2 - dimensional sphere (S^2/Z_2)

$$y_1^2 + y_2^2 \leq r_c^2.$$

At $y_2 = 0$ we have $-r_c \leq y_1 \leq r_c$, then the metric (15) will takes the form

$$ds^2 = e^{-2k|\phi|}g_{\mu\nu}dx^\mu dx^\nu - r_c^2 d\phi^2, \quad (16)$$

in which $y_1 = r_c\phi, -1 \leq \phi \leq 1$.

Metric (16) is exactly the RS metric if we identify (16) at $\phi = 1$ with that at $\phi = -1$. Hence, the manifold (15) contains two RS space - time (16) as its submanifolds at $y_1 = 0$ and $y_2 = 0$.

(ii) Compactification on a square (T^2/Z_2)

$$0 \leq y_1, y_2 \leq L.$$

In this case (15) is also to be reduced to (16) at $y_1 = 0, L$ and $y_2 = 0, L$. As a result, the manifold (15) contains two RS manifolds at the square summits if we identify the point $y_1 = 0$ with $y_1 = L$ and the point at $y_2 = 0$ with $y_2 = L$.

Journal of Physics: Conference Series **627** (2015) 012007 doi:10.1088/1742-6596/627/1/012007

3. The Planck mass and hierarchy problem

We consider scalar field $T(x)$ so that its VEV is the compact radius $r_c = const$. The local 4 - dimensional metric is given by

$$g_{\mu\nu} = \eta_{\mu\nu} + h_{\mu\nu}, \tag{17}$$

where $h_{\mu\nu}$ is symmetric tensor and very small. It is identified to physical gravitational field $h_{\mu\nu}$ of 4 - dimensional effective theory, $|h_{\mu\nu}| \ll 1$.

The 6 - dim metric takes the form

$$ds^2 = e^{-2k|T(x)|}\left[\eta_{\mu\nu} + h_{\mu\nu}\right] dx^\mu dx^\nu - dT^2(x) - T^2(x)d\theta^2. \tag{18}$$

The four dimensional effective action is given by

$$S_{eff} = \int d^4x \int dT(x) \int r_c d\theta \sqrt{-\bar{g}} e^{-4k|T(x)|}\left[-\Lambda + 2M^4 e^{2k|T(x)|}\bar{R}\right], \tag{19}$$

where $\bar{g} = det g_{\mu\nu}, \bar{R} = Re^{-2k|T(x)|}$.

We focus on the mass term in the 4 - dim gravitational action

$$S_{grav} = -\int d^4x \sqrt{-\bar{g}} 2M_{Pl}^2 \bar{R}, \tag{20}$$

where M_{Pl} is 4 - dimensional Planck mass.

Compare (19) to (20), we derive

$$M_{Pl}^2 = -M^4 \int_{-\pi}^{\pi} r_c d\theta \int e^{-2k|T(x)|} dT(x) = \frac{M^4 r_c}{2k} \int_{-\pi}^{\pi} d\theta e^{-2k|T(x)|}. \tag{21}$$

With $|T(x)| = r_c|\theta|$, it becomes

$$M_{Pl}^2 = \frac{M^4}{2k^2}\left(1 - e^{-2kr_c\pi}\right). \tag{22}$$

Similarly in the case of 5 dimensions, the result (22) shows that M_{Pl} depends weakly on r_c in the large kr_c limit. By means an adequate value of k, the 6 - dimensional mass M^4 has the same power with 4 - dimensional Planck mass. That means the hierarchy problem could be solved.

4. The Schwarzchild solution exptended in the 6D space-time

In this section we find the tensor metric that satisfies the central symmetry in 6D extended Einstein equation in the vacuum

$$R_N^M - \frac{1}{2}\delta_N^M R = -\Lambda \delta_N^M, \tag{23}$$

where we denote the Latin labels $M, N = 0, 1, 2, 3, 5, 6$ with $x^0 = t, x^1 = r, x^2 = \theta, x^3 = \varphi, x^5 = \rho, x^6 = \phi$. The metric in the spherical coordinates system corresponding to (23) reads

$$ds^2 = dt^2 - \left[dr^2 + r^2\left(d\theta^2 + \sin^2\theta d\varphi^2\right)\right] - d\rho^2 - \rho^2 d\phi^2. \tag{24}$$

Based on the Schwarzschild solution in 4D, we impose the metric in the form

$$G_{MN} = diag\left(e^{2\alpha(r)}, -e^{2\beta(r)}, -r^2, -r^2\sin^2\theta, -1, -\rho^2 e^{2\sigma(r)}\right). \tag{25}$$

Tensor Ricci R_N^M and R in the Eq (23) are calculated from the non vanishing Christoffel symbols Γ_{MN}^P. In two external dimensions it is determined by

$$\Gamma_{66}^5 = -\left\{\rho^2\sigma'(\rho) + \rho\right\} e^{2\sigma(\rho)}; \qquad \Gamma_{56}^6 = \Gamma_{65}^6 = \sigma'(\rho) + \frac{1}{\rho}; \tag{26}$$

$$R_{55} = \sigma'' + (\sigma')^2 + \frac{2\sigma'}{\rho}; \qquad R_{66} = \rho^2\left(\sigma'' + (\sigma')^2 + \frac{2\sigma'}{\rho}\right) e^{2\sigma(\rho)}. \tag{27}$$

The functions on the exponents of (25) are given by

$$\alpha(r) = -\beta(r) = ln\left(1 - \frac{C_1}{r}\right)^{\frac{1}{2}} \tag{28}$$

and

$$\sigma(\rho) = \ln\left(1 - \frac{C_2}{\rho}\right) \quad . \tag{29}$$

The constants C_1, C_2 are derived from the condition of the weak (Newton) gravitational field at the limit $r \to \infty$ and $\rho \to \infty$ (in the linear approximation)

$$C_1 = 2GM, C_2 = -GM. \tag{30}$$

Here Newton constant G is given by volumetric scaling of truly fundamental gravity scale G_*, which has dimension $[G_*] = [Energy]^{-(n+2)}$ [11].

$$G = \frac{G_*}{V_n}.$$

The extended Schwarzschild solution in 6D space-time takes the form

$$\begin{aligned} ds^2 = &\left(1 - \frac{2GM}{r}\right) dt^2 - \left(1 - \frac{2GM}{r}\right)^{-1} dr^2 \\ &- r^2\left(d\theta^2 + \sin^2\theta d\varphi^2\right) - d\rho^2 - \left(1 - \frac{2GM}{\rho}\right)^{-1} \rho^2 d\phi^2. \end{aligned} \tag{31}$$

Note that the metric (31) has singularities at $r = \rho = 2GM = r_0$, within which the spheres are trapped surfaces [12]. When $M \to 0$ the metric (31) is restored to 6D normal space-time (24). When $r \to \infty$ and $\rho \to 0$ this metric approaches the flat asymptotic.

5. The mass operator in extra dimensions

We have already determined the bulk metric in 6 dimensions. Let us now discuss the equation like Klein-Gordon equation, from which the mass of particles in extra dimensions is presented.

We start from the Klein-Gordon in the $(1, 3)$ dimensions

$$\left(\Box - m^2\right)\varphi(x_\mu) = 0. \tag{32}$$

Note that the spinor field which is described by the Dirac equation

$$i\gamma^\mu \frac{d\psi}{dx^\mu} + m\psi = 0,$$

also satisfies the Klein-Gordon equation. Hence, equation (32) could also determine the mass of fermions.

In the $(1, 3+n)$ dimensions, this equation takes the general form

$$\left(\Box + \sum_{a=1}^{n} \frac{\partial^2}{\partial y_a^2} - M^2\right) \Phi(x_\mu, y_a) = 0. \tag{33}$$

By separating the variables

$$\Phi(x_\mu, y_a) = \varphi(x_\mu)\psi(y_a),$$

we have

$$\begin{aligned} \left(\Box + \sum_{a=1}^{n} \frac{\partial^2}{\partial y_a^2} - M^2\right) \varphi(x_\mu)\psi(y_a) &= 0, \\ \frac{\Box\varphi(x_\mu)}{\varphi(x_\mu)} = -\frac{\sum_{a=1}^{n}\left(\frac{\partial^2}{\partial y_a^2} - M^2\right)\psi(y_a)}{\psi(y_a)} &= m^2. \end{aligned} \tag{34}$$

Hence we derive two equations

$$\left(\Box - m^2\right)\varphi(x_\mu) = 0, \tag{35}$$

$$\left(M^2 - \sum_{a=1}^{n} \frac{\partial^2}{\partial y_a^2}\right)\psi(y_a) = m^2\psi(y_a). \tag{36}$$

Equation (35) expresses that m is the mass of particles in physical space-time and equation (36) shows that m is the eigenvalue of spectrum mass operator corresponding to eigenfunction $\psi(y_a)$ in extra dimensions.

$$\widehat{m^2} = M^2 - \sum_{a=1}^{n} \frac{\partial^2}{\partial y_a^2}. \tag{37}$$

By transfering to the momentum space

$$\psi(y_a) = \int \frac{d^n y}{(2\pi)^{n/2}} u(q_a) \exp(ip^a y_a), \tag{38}$$

we obtain

$$m^2 = M^2 - \sum_{a=1}^{n} p_a^2 \quad \text{or} \quad \sum_{a=1}^{n} p_a^2 = m^2 - M^2. \tag{39}$$

It shows that the extra momenta are compactified on a sphere if $m^2 - M^2 > 0$. If $m^2 - M^2 < 0$ the extra momenta are compactified on a pseudo-sphere when one of the n extra dimensions is time-like. We are led to the conclusion that even in the models according to the universal extra-dimensions theory, where the extra dimensional space is not compactified, but the extra-momentum space must be compactified.

6. Discussion and conclusion

In the above sections, we have built a formalism for the warped six-dimensional space-time. It shows that this metric could be reduced to the RS space-time in the case of compactifications on two dimensional sphere (S^2/Z_2) or on a square (T^2/Z_2). Similarly in the case of RS space-time, the hierarchy problems is solved due to the Planck mass. In addition, the Schwarzschild solution in six dimensions is extended. We also proposed an equation for the eigenvalues of mass operator inclusive extra dimensions and showed that the extra momentum space must be compactified in any theory on extra dimensions.

It is hoped that this formalism could pave the way for better understanding of various physical phenomena of high energy physics.

Acknowledgments

This work is funded by Vietnam National Foundation for Science and Technology Development (NAFOSTED) under Grant 103.01-2011.50.

References

[1] Randall L and Sundrum R 1999 *Phys. Rev. Lett.* **83** 3370–3373
[2] Lim C S, Maru N, and Hasegawai K 2008 *J. Phys. Soc. Jap.* **77** 074101
[3] Dobrescu B A and Pontón E 2004 *J. High Energy Physics* (*Preprint* `arXiv:hepth/0401032`)
[4] Appelquist T, Dobrescu B A, Pontón E and Yee H U 2001 *Phys. Rev. Lett.* **87** 181802
[5] Appelquist T, Dobrescu B A, Pontón E and Yee H U 2002 *Phys. Rev.* D **65** 105019
[6] Cheng H C, Dobrescu B A and Hill C T 2000 *Nucl. Phys.* B 249–268
[7] Dobrescu B A and Poppitz E 2001 *Phys. Rev. Lett.* **87** 031801
[8] Hall L, Nomura Y, Okui T and Smith D 2002 *Phys. Rev.* D **65** 035008
[9] Hebecker A and March-Rusell J 2002 *Nucl. Phys. B* B625 128–150
[10] Bayntun A, Burgess C and van Nierop L 2010 *New J. of Phys.* **12** 075015
[11] Pérez-Lozenzana A 2005 *J. Phys. Conf. Ser. 18.* 224–269
[12] Weinberg S 1972 *Gravitation and Cosmology: Principles and Applications of the General Theory of Relativity* (Wiley)

Journal of Physics: Conference Series **627** (2015) 012008 doi:10.1088/1742-6596/627/1/012008

Some phenomenological studies of a $U(1)_R-$ lepton number model with a right handed neutrino

Sabyasachi Chakraborty

Department of Theoretical Physics, Indian Association for the Cultivation of Science, 2A & 2B Raja S. C. Mullick Road, Jadavpur, Kolkata 700 032, India.

E-mail: tpsc3@iacs.res.in

Abstract. We discuss the Higgs sector and neutrino mass and mixing in the context of a supersymmetric model with a $U(1)_R$ symmetry augmented by a single right handed neutrino superfield. The lepton number of the standard model fermions are identified with the negative of their R-charges. The presence of the neutrino Yukawa coupling (f) plays a very interesting role in Dirac neutrino mass generation as well as providing a tree level correction to the Higgs boson mass, when $f \sim \mathcal{O}(1)$. In order to fit the complete neutrino mass and mixing, a small breaking of the $U(1)_R$ symmetry is considered in the context of anomaly mediated supersymmetry breaking. We also show that when $f \sim 10^{-4}$, the right handed neutrino can be identified with a 7 keV sterile neutrino dark matter as recent studies have speculated.

1. Introduction

Perhaps the most important discovery in high energy physics in recent times is the observation of a neutral boson by the two LHC based experiments CMS [1] and ATLAS [2], widely speculated to be the Higgs boson with mass $\sim$ 125 GeV. All the decay channels of this boson is measured with moderate accuracy [3, 4, 5, 6, 7, 8, 9, 10, 11] and the signal strengths corresponding to different Higgs boson decay mode are found to be well in agreement with the standard model (SM) expectations. These results have constrained many beyond standard model (BSM) scenarios. However indirect hints of new physics cannot be ruled out either. In this regard supersymmetry remains one of the most celebrated BSM candidates although the two LHC based experiments have found no excess over the standard model background. These studies have put severe constraints on the superpartner masses. In this light R-symmetric models with Dirac gauginos are well motivated since they can explain the observation of a 125 GeV Higgs boson and provide a suitable dark matter candidate. Neutrino masses and mixing can be explained at the tree as well as at the one loop level. Finally and most importantly the presence of Dirac gluinos can also relax the bound on the first two generations of squark masses.

Motivated by these results we considered a $U(1)_R$ model [12, 13, 14] and extended it with a single right handed neutrino superfield [15]. The R-charges are identified with the negative of their lepton numbers. An important outcome of having R-symmetry is the absence of Majorana gaugino masses. However, Dirac masses for the gauginos are possible and in order to construct such a Dirac mass term, one must include chiral superfields, living in the adjoint representation of the SM gauge group. For example a singlet $\hat{S}$, a triplet $\hat{T}$ under $SU(2)_L$ and an octet $\hat{O}$ under $SU(3)_c$ help to formulate Dirac gaugino masses. The main artifact of this extension with a right handed neutrino superfield is to have a Dirac neutrino mass at the tree level itself. With some tuning of the parameters it is quite easy to show that even in the presence of a large neutrino Yukawa coupling $f \sim \mathcal{O}(1)$, a suitable small Dirac neutrino mass can be constructed. However a large f has other advantages as well. It provides a large tree level contribution [15] to the Higgs boson mass, thereby requiring negligible radiative corrections.

The R-symmetric case is interesting and should be explored further. Although, in the context of our work we consider the R-symmetry to be broken mildly. Gravitino mass happens to be the order parameter of R-breaking in the case of anomaly mediation. The neutrino Yukawa coupling plays a very interesting role and leads to novel phenomenology. For example large f scenario happens to have a bino like neutralino which is the next to lightest neutralino. The lightest neutralino turns out to be a Majorana neutrino with mass around 0.1 eV. Since the Higgs boson can decay to a pair of light neutralinos therefore the presence of this light bino like state enhances the total decay width of the Higgs boson, which constrain the parameter space severely and put important bounds on certain parameters. In addition, the small $f \sim \mathcal{O}(10^{-4})$ is also important as because it provides a sterile neutrino eigenstate which can be accommodated as the 7 keV dark matter [16] with appropriate active-sterile mixing.

2. $U(1)_R-$ lepton number model with a right handed neutrino

Along with the standard MSSM superfields $\hat{H}_u$, $\hat{H}_d$, $\hat{Q}_i$, $\hat{U}^c_i$, $\hat{D}^c_i$, $\hat{L}_i$, $\hat{E}^c_i$, the model includes a pair of vector like inert $SU(2)_L$ doublet superfields. Chiral superfields $\hat{S}$, $\hat{T}$ and $\hat{O}$ are required to construct 'supersoft' Dirac gaugino mass terms [17]. In addition we also have the right handed neutrino superfield $\hat{N}^c$. The R-charge assignments of the superfields are as follows,

	$\hat{Q}_i$	$\hat{U}^c_i$	$\hat{D}^c_i$	$\hat{L}_i$	$\hat{E}^c_i$	$\hat{H}_u$	$\hat{H}_d$	$\hat{R}_u$	$\hat{R}_d$	$\hat{S}$	$\hat{T}$	$\hat{O}$	$\hat{N}^c$
$U(1)_R$	1	1	1	0	2	0	0	2	2	0	0	0	2

Table 1. $U(1)_R$ charge assignments of the chiral superfields.

In a generic R-symmetric model the scalars possess the same R-charge as that of the superfields, whereas the fermions acquire R-charge one less than that of the superfields. We identify the lepton number as the negative of their R-charges [12]. The R-charge assignments from table 1 implies the left handed sneutrino vacuum expectation values (vevs) do not violate lepton number. Therefore, these vevs can be large and are not constrained from the neutrino Majorana mass. In order to have an invariant action under R-symmetry, the superpotential must carry R-charge of two units. Without any loss of generality the generic superpotential can be simplified by choosing a suitable basis in which one of the sneutrinos acquire a vev, in this case we choose it to be the electron type sneutrino. In addition, both $\hat{R}_u$ and $\hat{H}_d$ fields can be integrated once the coefficient of the bilinear term $\hat{H}_u\hat{R}_d$ is considered to be very large. The sneutrino can then play the role of a down type Higgs field. With all these assumptions in place, the superpotential is of the form

$$W = y^u_{ij}\hat{H}_u\hat{Q}_i\hat{U}^c_j + \mu_u\hat{H}_u\hat{R}_d + f\hat{L}_a\hat{H}_u\hat{N}^c + \lambda_S\hat{S}\hat{H}_u\hat{R}_d + 2\lambda_T\hat{H}_u\hat{T}\hat{R}_d - M_R\hat{N}^c\hat{S} + W', \quad (1)$$

and

$$W' = \sum_{b=2,3} f^l_b\hat{L}_a\hat{L}_b\hat{E}^c_b + \sum_{k=1,2,3} f^d_k\hat{L}_a\hat{Q}_k\hat{D}^c_k + \sum_{k=1,2,3}\frac{1}{2}\lambda_{23k}\hat{L}_2\hat{L}_3\hat{E}^c_k + \sum_{j,k=1,2,3;b=2,3}\lambda_{bjk}\hat{L}_b\hat{Q}_j\hat{D}^c_k, \quad (2)$$

where $a = e$ and $b = \mu, \tau$. One can always choose a basis in which the scalar component of the superfield $\hat{L}_a$ acquires a non zero *vev* (i.e. $< \tilde{\nu} >\equiv v_a \neq 0$) and the other two sneutrinos do not get any *vev*. We also note from eq. (2) that such an R-charge assignment enables us to write down a subset of lepton number preserving but trilinear R-parity violating terms in the superpotential.

The soft supersymmetry breaking but $U(1)_R$ preserving terms in the scalar sector can be generated by a spurion superfield $\hat{X}$, defined as $\hat{X} = x + \theta^2 F_X$. A nonzero vev of the F term generates soft supersymmetry breaking terms as written,

$$\begin{aligned} V_{soft} &= m^2_{H_u} H^\dagger_u H_u + m^2_{R_d} R^\dagger_d R_d + m^2_{\tilde{L}_a} \tilde{L}^\dagger_a \tilde{L}_a + \sum_{b=2,3} m^2_{\tilde{L}_b} \tilde{L}^\dagger_b \tilde{L}_b + M^2_N \tilde{N}^{c\dagger} \tilde{N}^c + m^2_{\tilde{R}_i} \tilde{l}^\dagger_{Ri} \tilde{l}_{Ri} \\ &+ m^2_S S^\dagger S + 2m^2_T \mathrm{tr}(T^\dagger T) + 2m^2_O \mathrm{tr}(O^\dagger O) - (b\mu_L H_u \tilde{L}_a + \mathrm{h.c.}) + (t_S S + \mathrm{h.c.}) \\ &+ \frac{1}{2} b_S (S^2 + \mathrm{h.c.}) + b_T(\mathrm{tr}(TT) + \mathrm{h.c.}) + B_O(\mathrm{tr}(OO) + \mathrm{h.c.}). \end{aligned} \tag{3}$$

Finally, the Dirac gaugino masses are generated by a spurion superfield $W'_\alpha = \lambda_\alpha D'$ where the supersymmetry breaking if of the D-type. The relevant Lagrangian takes the form [18],

$$\mathcal{L}^{\mathrm{Dirac}}_{\mathrm{gaugino}} = \int d^2\theta \frac{W'_\alpha}{\Lambda} [\sqrt{2}\kappa_1 \; W_{1\alpha} \hat{S} + 2\sqrt{2}\kappa_2 \; \mathrm{tr}(W_{2\alpha} \hat{T}) + 2\sqrt{2}\kappa_3 \; \mathrm{tr}(W_{3\alpha} \hat{O})] + h.c. \tag{4}$$

3. Scalar sector

In the R-symmetric scenario the CP even scalar squared mass matrix written in the basis $(h_R, \tilde{\nu}_R, S_R, T_R)$ is

$$\begin{aligned} (M^2_S)_{11} &= \frac{(g^2+g'^2)}{2} v^2 \sin^2\beta + (fM_R v_S - b\mu^a_L)(\tan\beta)^{-1} + 2\delta\lambda_u v^2 \sin^2\beta, \\ (M^2_S)_{12} &= f^2 v^2 \sin 2\beta + b\mu^a_L - \frac{(g^2+g'^2-2\delta\lambda_3)}{4} v^2 \sin 2\beta - fM_R v_S, \\ (M^2_S)_{13} &= 2\lambda^2_S v_S v \sin\beta + 2\mu_u \lambda_S v \sin\beta + 2\lambda_S \lambda_T v v_T \sin\beta + \sqrt{2} g' M^D_1 v \sin\beta - fM_R v \cos\beta, \\ (M^2_S)_{14} &= 2\lambda^2_T v_T v \sin\beta + 2\mu_u \lambda_T v \sin\beta + 2\lambda_S \lambda_T v_S v \sin\beta - \sqrt{2} g M^D_2 v \sin\beta, \\ (M^2_S)_{22} &= \frac{(g^2+g'^2)}{2} v^2 \cos^2\beta + (fM_R v_S - b\mu^a_L)\tan\beta + 2\delta\lambda_\nu v^2 \cos^2\beta, \\ (M^2_S)_{23} &= -\sqrt{2} g' M^D_1 v \cos\beta - fM_R v \sin\beta, \\ (M^2_S)_{24} &= \sqrt{2} g M^D_2 v \cos\beta, \\ (M^2_S)_{33} &= -\mu_u \lambda_S \frac{v^2 \sin^2\beta}{v_S} - \frac{\lambda_S \lambda_T v_T v^2 \sin^2\beta}{v_S} - \frac{t_S}{v_S} + \frac{g' M^D_1 v^2 \cos 2\beta}{\sqrt{2} v_S} + \frac{fM_R v^2 \sin 2\beta}{2 v_S}, \\ (M^2_S)_{34} &= \lambda_S \lambda_T v^2 \sin^2\beta, \\ (M^2_S)_{44} &= -\mu_u \lambda_T \frac{v^2}{v_T} \sin^2\beta - \lambda_S \lambda_T v_S \frac{v^2}{v_T} \sin^2\beta - \frac{g M^D_2}{\sqrt{2}} \frac{v^2}{v_T} \cos 2\beta, \end{aligned} \tag{5}$$

When the singlet and the triplet vev, v_S and v_T are small the corresponding scalar masses becomes very heavy. In such a scenario the scalar mass matrix can be essentially considered to be a 2×2 mass matrix. The lightest eigenvalue of the CP even scalar mass matrix can be identified with the 125 GeV Higgs boson. One very important outcome of this decoupling limit is the Higgs boson mass gets a contribution at the tree level which is of the following form [15],

$$M_h^2 \;=\; \Big[M_Z^2\cos^2 2\beta + f^2 v^2 \sin^2 2\beta\Big]. \tag{6}$$

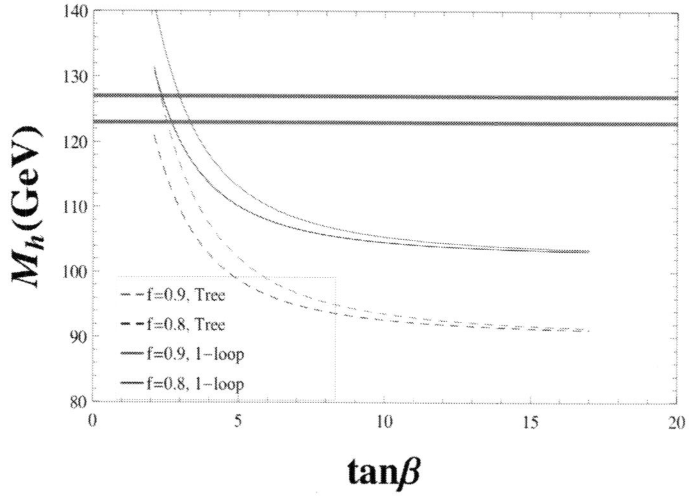

Figure 1. The dashed lines represent Higgs boson mass at the tree level only, whereas the continuous lines take into account radiative correction through stop loops with $M_{\tilde{t}} = 500$ GeV.

From eq. (6) it is conspicuous that the Higgs boson mass receives substantial contribution at the tree level itself for $f \sim \mathcal{O}(1)$ and for certain regions of $\tan\beta$. This statement is transparent from figure 1 that $M_h \sim 125$ GeV for order one Yukawa coupling and when $\tan\beta$ is small.

4. Neutrino sector: R preserving case

In the sector of neutral fermions we have a mixing between neutrino, neutral gauginos and neutral higgsinos, i.e. the electron type neutrino ν_e mixes with the sterile neutrino N^c. The neutralino mass matrix written in the basis $\psi^{0+} = (\tilde{b}^0, \tilde{w}^0, \tilde{R}_d^0, N^c)$ and $\psi^{0-} = (\tilde{S}, \tilde{T}^0, \tilde{H}_u^0, \nu_e)$

$$M_\chi^D = \begin{pmatrix} M_1^D & 0 & \frac{g' v_u}{\sqrt{2}} & -\frac{g' v_a}{\sqrt{2}} \\ 0 & M_2^D & -\frac{g v_u}{\sqrt{2}} & \frac{g v_a}{\sqrt{2}} \\ \lambda_S v_u & \lambda_T v_u & \mu_u + \lambda_S v_S + \lambda_T v_T & 0 \\ M_R & 0 & -f v_a & -f v_u \end{pmatrix}. \tag{7}$$

The lightest eigenvalue of this neutralino mass matrix corresponds to the Dirac neutrino mass which we deduce as [15],

$$m_{\nu_e}^D \;=\; \frac{v^3 f g \sin\beta}{\sqrt{2}\gamma M_1^D M_2^D}\lambda_T (M_2^D - M_1^D), \tag{8}$$

where we assumed the following relations, $\lambda_T = \lambda_S \tan\theta_W$ and $M_R = \frac{\sqrt{2} f M_1^D \tan\beta}{g \tan\theta_W}$. We observe from eq. (8) that even in the presence of $f \sim \mathcal{O}(1)$, Dirac neutrino mass can be appropriately small for small $\lambda_T \sim 10^{-6}$ and nearly degenerate Dirac gaugino masses.

Journal of Physics: Conference Series **627** (2015) 012008 doi:10.1088/1742-6596/627/1/012008

5. Neutrino sector: R breaking case

Recent cosmological observations predict a vanishingly small value for the cosmological constant. To explain this in the framework of a spontaneously broken supergravity theory, the superpotential has to acquire a non-zero vev at the vacuum. The gravitino mass is also proportional to the vev of the superpotential in the vacuum, therefore the gravitino can be thought of as the order parameter of R-breaking. The information of the breaking of R-symmetry would have to be transferred from the hidden sector to the visible sector and we assume anomaly mediation (AMSB) playing the role of this messenger. The breaking of R-symmetry would generate Majorana gaugino masses and trilinear scalar couplings. In the R-breaking scenario the neutralino mass matrix written in the basis $(\widetilde{b}^0, \widetilde{S}, \widetilde{w}^0, \widetilde{T}, \widetilde{R}_d^0, \widetilde{H}_u^0, N^c, \nu_e)$ is given as,

$$M_\chi^M = \begin{pmatrix} M_1 & M_1^D & 0 & 0 & 0 & \frac{g' v_u}{\sqrt{2}} & 0 & -\frac{g' v_a}{\sqrt{2}} \\ M_1^D & 0 & 0 & 0 & \lambda_S v_u & 0 & M_R & 0 \\ 0 & 0 & M_2 & M_2^D & 0 & -\frac{g v_u}{\sqrt{2}} & 0 & \frac{g v_a}{\sqrt{2}} \\ 0 & 0 & M_2^D & 0 & \lambda_T v_u & 0 & 0 & 0 \\ 0 & \lambda_S v_u & 0 & \lambda_T v_u & 0 & \mu_u + \lambda_S v_S + \lambda_T v_T & 0 & 0 \\ \frac{g' v_u}{\sqrt{2}} & 0 & -\frac{g v_u}{\sqrt{2}} & 0 & \mu_u + \lambda_S v_S + \lambda_T v_T & 0 & -f v_a & 0 \\ 0 & M_R & 0 & 0 & 0 & -f v_a & 0 & -f v_u \\ -\frac{g' v_a}{\sqrt{2}} & 0 & \frac{g v_a}{\sqrt{2}} & 0 & 0 & 0 & -f v_u & 0 \end{pmatrix}. \tag{9}$$

The lightest neutralino corresponds to the Majorana neutrino which has a mass as,

$$(m_\nu)_{\rm Tree} \simeq -v^2 \frac{\left[g \lambda_T v^2 (M_2^D - M_1^D) \sin\beta\right]^2}{\left[M_1 \alpha^2 + M_2 \delta^2\right]}, \tag{10}$$

where

$$\begin{aligned} \alpha &= \frac{2 M_1^D M_2^D (\mu_u + \lambda_S v_S + \lambda_T v_T) \tan\beta}{g \tan\theta_W} + \sqrt{2} v^2 \lambda_S \tan\beta (M_1^D \sin^2\beta + M_2^D \cos^2\beta), \\ \delta &= \sqrt{2} M_1^D v^2 \lambda_T \tan\beta. \end{aligned} \tag{11}$$

The full neutrino mass and mixing can be explained once one loop radiative corrections are taken into consideration as shown in figure 2.[1]. The characteristics of the next to lightest neutralino depend on the magnitude of the neutrino Yukawa coupling. Let us now discuss these two cases separately.

5.1. *Case with small neutrino Yukawa coupling: $f \sim \mathcal{O}(10^{-4})$*

In the limit when $f \sim \mathcal{O}(10^{-4})$ the Majorana neutrino mass is given in eq. (10). The sterile neutrino mass computed as $M_N^R \simeq M_1 \frac{2 f^2 \tan^2\beta}{g'^2}$. The active-sterile mixing is estimated as $\theta_{14}^2 \simeq \frac{(m_\nu)_{\rm Tree}}{M_N^R}$ [16]. With these analytical expressions in hand, we fit a 7 keV sterile neutrino dark matter explaining the emergence of a weak line at 3.5 keV in the X-ray spectra of the Andromeda galaxy and various other galaxy clusters including the Perseus galaxy cluster as observed by XMM-Newton X-ray space observatory [19, 20]. The observed flux and the peak of the energy

[1] For detailed calculations see [14, 15]

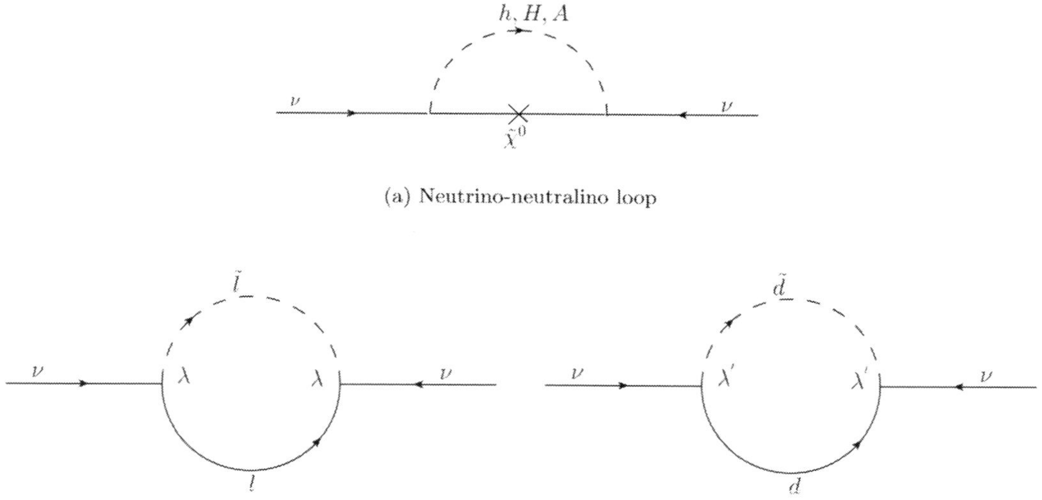

(a) Neutrino-neutralino loop

(b) Lepton-slepton loop

(c) Quark-squark loop

Figure 2. One loop diagrams to generate neutrino masses.

can be translated to an active-sterile mixing in the range $2.2 \times 10^{-11} < \sin^2 2\theta_{14} < 2 \times 10^{-10}$. The mass and mixing can be easily conceived in our model. From figure 3 we observe that the Dodelson-Widrow (DW) [21] mechanism of sterile neutrino production is not compatible with such a small active-sterile neutrino mixing as pointed out in refs. [19, 20]. However, resonant production of the sterile neutrinos [22, 23] in the presence of a cosmological lepton asymmetry $L \sim \mathcal{O}(10^{-3})$ is capable of producing correct relic density of 0.119 [24].

Further, such a precise measurement of the mass and the mixing of this sterile neutrino provides substantial bound on the model parameters. For example the maximum value of the active-sterile mixing as computed from the recent energy peak at 3.5 keV corresponds to the minimum of $\tan\beta$ and vice-versa. Therefore in figure 4 we show the variation of $\tan\beta$ as a function of M_2^D. We infer that the lower limit on $\tan\beta$ coming from the estimated mass and mixing of the sterile neutrino dark matter is stronger in a wide range of parameter space compared to the bound coming from electroweak precision measurement computations [12] as shown in figure 4. For higher values of M_2^D the upper limit on $\tan\beta$ becomes stronger than the upper bound coming from τ Yukawa coupling [12]. When combined these two limits provide a viable region for the Dirac wino mass.

Finally, it is important to check if this case [25] is consistent with the recent signal strengths, $\mu_{\gamma\gamma} = \frac{\sigma(pp\to h\to\gamma\gamma)}{\sigma(pp\to h\to\gamma\gamma)^{\rm SM}}$ obtained by both CMS and ATLAS experiments [2] or not. In this case the Higgs boson can decay to two light neutralinos namely the active and the sterile neutrino. The presence of light stops can provide certain enhancements in the signal strengths [25].

5.2. Case with large neutrino Yukawa coupling: $f \sim \mathcal{O}(1)$

As previously noted for the case of the large Yukawa coupling we obtain a bino like light neutralino with mass around a few hundred MeV, which increases the total decay width of

[2] We list the recent Higgs boson signal strengths in table 2

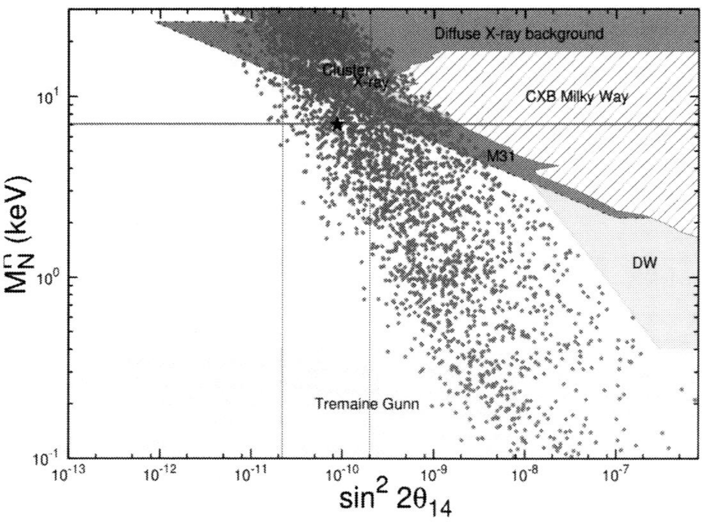

Figure 3. The red scattered points are obtained by varying the parameters in a wide range. All the shaded regions show different bounds obtained from different X-ray experiments [21, 27]. Finally the black star represents the central value of the observed mass and mixing by the XMM-Newton X-ray space observatory.

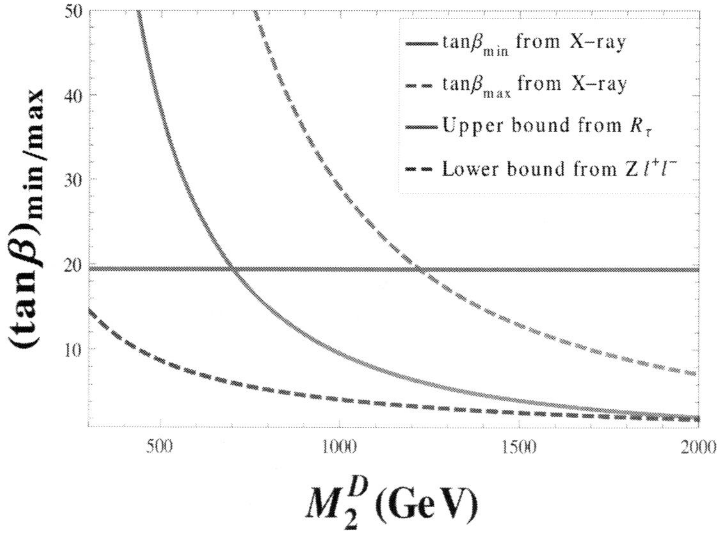

Figure 4. Showing the bound on $\tan\beta$ as a function of M_2^D.

the Higgs boson. The recent results from the CMS experiment shows a limit on the total decay width of Higgs as $k_{TOT} = \frac{\Gamma^{U(1)_R}_{h\to Total}}{\Gamma^{SM}_{h\to Total}} \leq 5.4$ [26]. This result could constrains our parameter space severely [25].

Channel	μ (CMS)	μ (ATLAS)
$h \rightarrow \gamma\gamma$	$1.14^{+\ 0.26}_{-\ 0.23}$ [3]	$1.17^{+\ 0.27}_{-\ 0.27}$ [4]
$h \xrightarrow{ZZ^*} 4l$	$0.93^{+\ 0.39}_{-\ 0.32}$ [5]	$1.44^{+\ 0.40}_{-0.33}$ [4]
$h \xrightarrow{WW^*} 2l2\nu$	$0.72^{+\ 0.20}_{-\ 0.18}$ [6]	$1.0^{+0.30}_{-0.30}$ [7]
$h \rightarrow b\bar{b}$	$1.0^{+\ 0.5}_{-\ 0.5}$ [8]	$0.2^{+\ 0.70}_{-\ 0.60}$ [9]
$h \rightarrow \tau\bar{\tau}$	$0.78^{+\ 0.27}_{-\ 0.27}$ [10]	$1.4^{+\ 0.5}_{-\ 0.4}$ [11]

Table 2. Signal strengths (μ) in different decay final states of the SM-like Higgs boson as reported by the CMS and the ATLAS collaborations (with the corresponding references).

6. Acknowledgments

I would like to thank Council of Scientific and Industrial Research (CSIR), Government of India for a Senior Research Fellowship. It is also a pleasure to thank Sourov Roy, Dilip Kumar Ghosh and AseshKrishna Datta for collaborations based on which this talk was presented. Finally I would also like to thank the organizers of 2nd International Workshop on Theoretical and Computational Physics (IWTCP-2), Buon Ma Thuot, Vietnam for giving me this opportunity to participate in the workshop.

References

[1] Chatrchyan S *et al.* 2012 *Phys. Lett.* B **716** 30
[2] Aad G *et al.* 2012 *Phys. Lett.* B **716** 1
[3] Khachatryan V *et al.* 2014 *Eur. Phys. J* C **74** 3076
[4] Aad G *et al.* (*Preprint* **hep-ex/1408.7084**)
[5] Chatrchyan S *et al.* 2014 *Phys. Rev.* D **89** 092007
[6] Chatrchyan S *et al.* 2014 *J. High Energy Phys.* JHEP01(2014)096
[7] ATLAS Collaboration 2013 *ATLAS-CONF-2013-034*
[8] Chatrchyan S *et al.* 2014 *Phys. Rev.* D **89** 012003
[9] ATLAS Collaboration 2013 *ATLAS-CONF-2013-079*
[10] Chatrchyan S *et al.* 2014 *J. High Energy Phys.* JHEP05(2014)104
[11] ATLAS Collaboration 2013 *ATLAS-CONF-2013-108*
[12] Frugiuele C, Gregoire T, Kumar P and Ponton E 2013 *J. High Energy Phys.* JHEP03(2013)156
[13] Frugiuele C, Gregoire T, Kumar P and Ponton E 2013 *J. High Energy Phys.* JHEP05(2013)012
[14] Bertuzzo E and Frugiuele C 2012 *J. High Energy Phys.* JHEP05(2012)100
[15] Chakraborty S and Roy S 2014 *J. High Energy Phys.* JHEP01(2014)101
[16] Chakraborty S, Ghosh D K and Roy S 2014 *J. High Energy Phys.* JHEP10(2014)146
[17] Fox P J, Nelson A E and Weiner N 2002 *J. High Energy Phys.* JHEP08(2002)035
[18] Benakli K, Goodsell M D and Maier A K 2011 *Nuclear Physics* B **851** 445, Benakli K 2011 *Fortsch. Phys.* **59** 1079
[19] Bulbul E, Markevitch M, Foster A, Smith R K, Loewenstein M and Randall S W 2014 *Astrophys. J.* **789** 13
[20] Boyarsky A, Ruchayskiy O, Iakubovskyi D and Franse J (*Preprint* **astro-ph/1402.4119**)
[21] Dodelson S and Widrow L M 1994 *Phys. Rev. Lett.* **72** 17
[22] Shi X D and Fuller G M 1999 *Phys. Rev. Lett.* **82** 2832
[23] Abazajian K, Fuller G M and Patel M 2001 *Phys. Rev.* D **64** 023501
[24] Abazajian K N 2014 *Phys. Rev. Lett.* **112** 161303
[25] Chakraborty S, Datta A and Roy S (*Preprint* **hep-ph/1411.1525**)
[26] Khachatryan V *et al.* 2014 *Phys. Lett.* B **736** 64
[27] Tremaine S and Gunn J E 1979 *Phys. Rev. Lett.* **42** 407, Boyarsky A, Ruchayskiy O and Iakubovskyi D 2009 *J. Cosmology and High Energy Phys.* JCAP03(2009)005, Abazajian K N, Markevitch M, Koushiappas S M and Hickox R C 2007 *Phys. Rev.* D **75** 063511, Watson C R, Beacom J F, Yuksel H and Walker T P 2006 *Phys. Rev.* D **74** 033009

Status and prospects of the Belle II experiment

Tomoyuki Konno

Department of Physics, Tokyo Metropolitan University, 1-1 Minami-oosawa, Hachioji-si, Tokyo, Japan

E-mail: konno@hepmail.phys.se.tmu.ac.jp

Abstract. Measurements using the Belle detector at the KEKB collider in Japan have achieved important progress in the flavor structure of the Standard Model of elementary particle physics. Both the collider and the detector are being upgraded to the SuperKEKB collider and the Belle II detector to collect about 40 times higher luminosity of the Belle experiment with several improvements in the detector performance. Using the higher statistics, several measurements with precision comparisons to the predictions in the Standard Model will be performed to search for new physics beyond the Standard Model from a wide range of flavor physics channels. This paper describes the physics prospects and collider/detector construction status of the Belle II experiments.

1. Introduction

The two previous B factory experiments using e^+–e^- asymmetric colliders, the Belle experiment [1] at the KEKB collider in KEK [2] and the BaBar experiment [3] at the PEP-II collider in SLAC, collected 1.5 ab^{-1} of integrated luminosity on $\Upsilon(4S)$ resonances, mainly decaying to $B\bar{B}$ meson pairs, in long term operation. Both colliders were designed with asymmetric beam energies of electron and positron to boost the B meson pair and the decay times of the $B\bar{B}$ mesons are measured from the lengths of the two particles. With the accumulated $B\bar{B}$ samples, some of important insights in the flavor structure of the Standard Model have been obtained. The observation of CP violation in the B meson system is one of the memorable results, confirming the structure of the quark flavor sector proposed by N. Cabibo, M. Kobayashi and T. Maskawa [4, 5]. However a lot of interesting measurements are still limited by statistics. To improve statistics, upgrades of both the KEKB collider and the Belle detector to the SuperKEKB collider [6] and the Belle II detector [7], are in progress in order to achieve 50 ab^{-1} of luminosity to search for physics beyond the Standard Model with more precise checks of the predictions of the Standard Model. This paper describes the status and physics prospects of the Belle II experiment in the SuperKEKB facility.

2. Physics prospects

A broad physics program in the heavy-flavor system with much higher statistics on the $\Upsilon(4S)$ resonance at the SuperKEKB collider and the Belle II detector is performed to search for signs of new physics phenomena beyond the Standard Model in the heavy-quark system [8, 9]. The quest is based on indirect searches for contributions from new physics, which are expected to show up as differences between measured values of the observables and the ones predicted in

Journal of Physics: Conference Series **627** (2015) 012009 doi:10.1088/1742-6596/627/1/012009

the Standard Model. Some important topics are selected among the large number of planned measurements and explained in the following paragraphs.

2.1. Search for non-Standard Model CP violation
The CP violation phase ϕ_1 in the Standard Model should be same among measurements of various decay modes while some new physics models such as supersymmetric grand-unified theories with the see-saw mechanism [10, 11, 12] allow deviations from the Standard Model predictions. One of the most promising ways is to compare the CP asymmetry in $b \to sq\bar{q}$, which is sensitive to possible contribution from the new physics models, with the asymmetry in $B \to J/\Psi K_S$. The measurements of $\sin\phi_1$ with $b \to sq\bar{q}$ are still statistically limited in the results of the Belle experiments and it is possible to reach the theoretical limit of 0.03, for the three golden modes ($B \to \phi K^0$, $B \to \eta' K^0$ and $B \to \phi K_S K_S K_S$), with 50 ab^{-1} of the integrated luminosity at the SuperKEKB collider and the Belle II detector [8].

2.2. $B \to \tau\nu$ measurements
$B \to \tau\nu$ decay is one of interesting channels for the new physics search since the branching fraction of $B \to \tau\nu$ is expected to be affected by the contribution from the charged Higgs in the so-called type II Higgs doublet models instead of the contribution of the weak boson in the Standard Model. The first evidence for the $B \to \tau\nu$ decay has been confirmed by the Belle experiment [13, 14], followed by the BaBar experiment at PEP-II [15, 16]. A small disagreement has been observed between the measured branching fraction of $B \to \tau\nu$ and the theoretical prediction in the Standard Model but these results are still consistent with the Standard Model prediction within errors. The Belle II experiment is expected to give a better constraint to the charged Higgs mass and to $\tan\beta$.

2.3. τ lepton flavor violation
The lepton-flavor-violating decays such as $\tau \to \mu\gamma$ could be induced by neutrino oscillation with non-zero neutrino mass but these processes are strongly suppressed due to the small neutrino mass and the branching fractions are far beyond experimental reach. However, the situation can differ if a new particle exists with a mass of the order of the weak scale and couples to leptons. Some extensions of the Standard Model predict enhanced lepton-flavor-violating decays and the branching fractions can be as high as experimentally accessible by the Belle II experiment, reaching at the order of 10^{-9} to 10^{-7} [17, 18].

3. SuperKEKB collider

The designed centre-of-mass energy for the SuperKEKB colider on the $\Upsilon(4S)$ resonance is the same as for the KEKB colider while the designed luminosity for the SuperKEKB colider is 8.0×10^{43} cm^{-2}s^{-1}, 40 times larger than 2.1×10^{34} achieved by the KEKB colider. Parameters of KEKB and SuperKEKB colliders are summarized in Table 1. The upgrade to the SuperKEKB collider is based on the "nano-beam" scheme proposed by P. Raimondi from Frascati [19]. The narrow beam size reduces the vertical beta function at the interaction point by a factor of 1/20, improving luminosity by a factor of 20. In order to achieve this improvement, the final focusing magnets are located closer to the interaction point enlarging the beam crossing angle form 22 mrad to 83 mrad. The other major contribution to the luminosity improvement is from the increase of the beam current which doubles the luminosity. The construction of the SuperKEKB facility is expected to be completed by the end of 2014 Japanese fiscal year, March 2015, and commissioning of the SuperKEKB colider will start in 2015.

Table 1. Parameters of the design values of SuperKEKB and the achieved values of KEKB. The two values separated by a slash indicate the values for positron and electron beams, respectively.

Parameters	SuperKEKB	KEKB
Beam energy (GeV)	4.0/7.0	3.5/8.0
Half crossing angle (mrad)	11	41.5
Vertical beta functions at IP (mm)	5.9	0/0.31
Beam currents (A)	1.64/1.19	3.60/2.60
Luminosity ($cm^{-2}s^{-1}$)	2.1×10^{34}	8×10^{35}

4. Belle II detector

The Belle detector is also being upgraded to the Belle II detector against an increase of background occupancy and radiation damages due to a rise in the beam related background, 20 times higher than the Belle experiment. The effects of the background in the detector performance are estimated with the results of the operation of the KEKB collider and the Belle detector by accounting for scaling for each component of the background. Furthermore, to exploit the physics potential of the Belle II experiment, identification capabilities of hadron and muon, especially with low momentum, are also considered in the detector design. Figure 1 shows the dimensions of the Belle II detector compared with the Belle detector. The Belle II detector is under construction and expected to start physics run in 2017. The status of each part of the Belle II detector are described bellow.

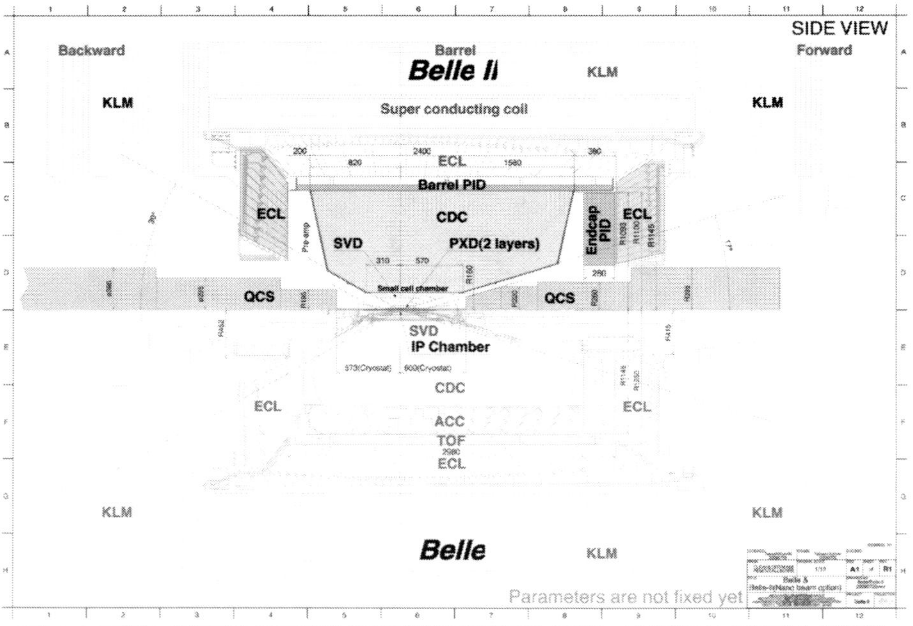

Figure 1. The Belle II detector (top half) compared with the Belle detector (bottom half).

Journal of Physics: Conference Series **627** (2015) 012009 doi:10.1088/1742-6596/627/1/012009

The four layer silicon strip detector is replaced by a two-layer silicon pixel detector (PXD) based on the DEPFET (DEpleted P-channel Field Effect Transistor) and a four layer silicon strip detector (SVD). The impact parameter resolution in the beam direction is improved mainly by the PXD by a factor of about 2 for momenta below 1 GeV and the improvement of the resolution is also contributed by bringing the detector closer to the interaction point. On the other hand, the SVD contributes in the improvement of reconstruction efficiency of K_S decays to two charged pions thanks to the larger outer radius coverage. Prototypes of the PXD and SVD were tested at an electron test beam facility of Deutsches Elektronen-Synchrotron (DESY) in January 2014. The prototypes were operated with the CO2 detector cooling system and the minimum set of the Belle II data acquisition system, resulting in adequate performance of vertex resolution and data readout with online data reduction.

The central drift chamber (CDC) has smaller drift cells near the beam pipe, extended outer radius coverage, and faster readout electronics compared with the one in the Belle detector to provide better performance of momentum and dE/dx resolution. The faster readout electronics achieves less dead time from several μs to 200 ns with higher event rate by adapting waveform sampling instead of charge-to-time converter.

The particle identification systems (PID) for charged pion and kaon in both the barrel and forward end cap regions of the Belle II detector are based on Cherenkov ring imaging detectors, upgraded from the Cherenkov threshold detectors. The probability of misidentifying a charged pion (kaon) as a charged kaon (pion) is improved from about 10 % to 1 % for selection efficiencies of about 95% at the maximum kinematic momentum of 4 GeV/c. Both PID systems reconstruct the Cherenkov radiation angles and identify the particle types from the differences in radiation angles. The barrel PID system, called TOP, is made of highly pure quartz radiator and two dimensional MCP-PMT, which measures both hit patterns and time-of-fight of the Cherenkov photons in the quartz material, internally reflected in the quartz module. On the other hand, the forward end cap PID system consists in aerogel radiators and HAPDs, called A-RICH, recording distribution of the Cherenkov photons as 2-D ring images on an array of HAPDs.

The electromagnetic calorimeter (ECL) is based on CsI crystal in the Belle detector and the Belle II detector reuses the material of the Belle detector in the barrel region while the material in the end cap regions is replaced by pure CsI crystal to increase radiation hardness and to improve performance against pileup problems. In addition to the replacement of the material, the readout electronics are also replaced to a 2MHz waveform sampling readout to suppress beam background contamination and improve energy resolution.

The outermost detector for the measurement of K_L and muon (KLM) in the Belle II detector is mainly based on reusing the RPCs of the Belle detector, partially replaced by plastic scintillator strips and optical fibers for the innermost two layers.

The increase of the beam luminosity causes much higher trigger rates up to 30 kHz at the first level trigger of the data acquisition (DAQ) system and the event data size also increases especially in the PXD to improve better performance in the high background environment with the finer granularity and more readout channels in the front end electronics. 30 GB/s of event data are read out in the PXD front end electronics and 3 GB/s are collected from the other detector electronics at the first level trigger. Several software codes for offline analysis are performed in the DAQ backend PC farm to generate the second level software trigger and the region of interest (RoI) on the PXD to reduce data size by squeezing limited data containing physics interests with reconstructed particle tracks. Finally, 3GB/s of event sample are stored on disk. The full readout chain from front end electronics to storage were tested at DESY during the beam test of the PXD and SVD prototypes in January 2014.

5. Summary

The Belle II detector at SuperKEKB is the upgrade of the previous Belle detector at the KEKB collider, expected to achieve about 40 times higher luminosity with several improvements of the detector performance. The Belle II experiment has a broad physics program to search for new physics beyond the Standard Model in the heavy flavor sector. The SuperKEKB facility is expected to complete the construction by the end of Japanese fiscal year 2014 and the accelerator commissioning will start in 2015 while the construction of the Belle II detector is now ongoing and expected to start taking physics data in 2017.

References

[1] Abashian A *et al.* (Belle Collaboration) 2002 *Nucl. Instrum. Methods Phys. Res.* A **479** 117-232
[2] Kurokawa S and Kikutani E 2003 *Nucl. Instrum. Methods Phys. Res.* A **499** 1-7
[3] Aubert B et al. (BABAR Collaboration) 2002 *Nucl. Instrum. Meth.* A **479** 1-116
[4] Cabibbo N 1963 *Phys. Rev. Lett.* **10** 531
[5] Kobayashi M and Maskawa T 1973 *Prog. Theor. Phys.* **49** 652
[6] Golob B 2010 *Nuovo Cim.* **C33** 319-326
[7] Abe T *et al.* 2010 *KEK Report 2010-1* [arXiv:1011.0352]
[8] AushevT *et al.* 2009 *KEK Report 2009-12* [arXiv:1002.5012]
[9] Isidori G *et al.* 2010 *Ann. Rev. Nucl. Part. Sci.* **60** 355
[10] Mori T 2000 *Phys. Lett.* B **493** 366
[11] Baek TS, Goto T, Okada Y Okumura K 2001 *Phys. Rev.* D **63**, 051701
[12] Chang D, Masiero A and Murayama H 2003 *Phys. Rev.* D **67**, 075013
[13] Hara K *et al.* (Belle Collaboration) 2013 *Phys. Rev. Lett.* **110**131801
[14] Hara K *et al.* (Belle Collaboration) 2010 *Phys. Rev.* D **82** 071101(R)
[15] Lees J. P *et al.* (BABAR Collaboration) 2013 *Phys. Rev.* D **88**, 031102
[16] B. Aubert *et al.* (BABAR Collaboration) 2010 *Phys. Rev.* D **81** 051101
[17] Ellis J, Hisano J, Raidal M and Shimizu Y 2002 *Phys. Rev.* D **66** 115013
[18] Brignole A and Rossi A 2004 *Nucl. Phys.* B **701** 3
[19] Milardi C *et al.* 2009 *ICFA Beam Dyn. Newslett* **48** 23-33

Journal of Physics: Conference Series **627** (2015) 012010 doi:10.1088/1742-6596/627/1/012010

Handling accuracy in Bayesian automatic adaptive quadrature

Gh Adam[1,2] **and S Adam**[1,2]

[1] Laboratory of Information Technologies, Joint Institute for Nuclear Research, 6 Joliot Curie St., 141980 Dubna, Moscow Region, Russia

[2] Horia Hulubei National Institute for Physics and Nuclear Engineering (IFIN-HH), 30 Reactorului St., Magurele - Bucharest, 077125, Romania

E-mail: adamg@jinr.ru

Abstract. New Bayesian inferences which significantly extend the coverage of the Bayesian automatic adaptive quadrature (BAAQ) approach to the solution of Riemann integrals are reported. The scrutiny of the possible floating-point machine number approximations of abscissa values inside an integration domain unveiled the occurrence of five classes of integration domains entering the quadrature problems: zero-length, open void, microscopic, mesoscopic, macroscopic. Correct approach to the class identifications and class adapted advancement to the solution are described. In the most complex, macroscopic case, the reversible addition of a new decision branch extends the BAAQ approach coverage to both difficult and easy integrals. The reliability of the code used for solving easy integrals was enhanced with four new analysis tools as compared to the standard automatic adaptive quadrature solution.

1. Introduction

We describe new progress toward the derivation and implementation of the *Bayesian automatic adaptive quadrature* (BAAQ) [1, 2] as a *robust, reliable, fast,* and *highly accurate* computational tool of interest to the *modeling of physical phenomena* within *numerical experiments* asking for the evaluation of large numbers of *Riemann integrals* by numerical methods. Instances of such phenomena are, e.g., the phase transitions or processes involving fragmentation and fusion, where the behaviour of the system is characterized by the *sudden change* of an *inner order parameter*, which results in drastic and unpredictable modification of the mathematical properties of the integrands.

Such circumstances cannot be accommodated within the *standard automatic adaptive quadrature* (AAQ) approach to the numerical solution (see, e.g., dedicated chapters in the monographs [3, 4, 5]), due to the impossibility to decide in advance on the right choice of the convenient library procedure from an existing menu (see, e.g., [6]).

The presentation is structured in five sections: pros and cons of the standard AAQ (section 2), Bayesian inferences on the classes of integration domain lengths (section 3), integration domain adapted BAAQ algorithms (section 4), and conclusions (section 5).

2. Pros and cons of the automatic adaptive quadrature

2.1. The mathematical problem

We seek a numerical solution of the (proper or improper) Riemann integral

$$I \equiv I[a,b]f = \int_a^b w(x)f(x)dx\,, \tag{1}$$

over an integration domain $[a,b] \subseteq \mathbf{R}$, for a real valued *integrand function* $f : [a,b] \to \mathbf{R}$ which is almost continuous everywhere on $[a,b]$, such that (1) exists and is finite. The *weight function* $w(x)$ either absorbs an analytically integrable difficult factor entering the integrand (e.g., an oscillatory or endpoint singular function), or else $w(x) \equiv 1$, $\forall x \in [a,b]$.

2.2. Numerical quadrature

The numerical solution of (1) consists of a pair $\{Q, E > 0\}$, where Q is an *approximate value* of I, while E is an estimate of the *error associated to* Q. The obtained output is asked to satisfy the accuracy requirement

$$|I - Q| < E < \tau\,, \tag{2}$$

where the prescribed input accuracy τ is defined in terms of two parameters: the absolute accuracy ε_a and the relative accuracy ε_r, such that [3]

$$\tau = \max\{\varepsilon_r|I|, \varepsilon_a\} \simeq \max\{\varepsilon_r|Q|, \varepsilon_a\}\,. \tag{3}$$

If the condition $E < \tau$ is not fulfilled, the value of Q is refined and a new error estimate E is computed. If this process is converging, then (2) is eventually fulfilled and the computation is terminated.

2.3. Permanent features of the automatic adaptive quadrature

The computation scheme developed within the AAQ approach implements an *integrand adapted* discretization of $[a,b]$, which defines a *partition* of $[a,b]$,

$$\Pi_N[a,b] \equiv \{a = x^0 < x^1 < \cdots < x^i < \cdots < x^N = b|\ N \geq 1\}\,. \tag{4}$$

Over each subrange $[x^{i-1}, x^i] \subseteq [a,b]$, a (possibly subrange dependent) *local quadrature rule* is used to get an approximate value q of $I[x^{i-1}, x^i]f$, together with an associated error estimate $e > 0$.

A partition dependent global pair solving (1), $\{Q \equiv Q_N, E \equiv E_N > 0\}$, is got by summing up the individual outputs $\{q, e > 0\}$ over the subranges of the partition (4). In what follows, in order to avoid cumbersome notations, we will consider a *generic subrange* $[\alpha, \beta] \subseteq [a,b]$, standing for any subrange $[x^{i-1}, x^i]$ of (4).

The number of subranges of (4) starts with $N = 1$ and it is increased by gradual refinement of $\Pi_N[a,b]$ until either the accuracy condition (2)–(3) is fulfilled, or a failure diagnostic is issued.

2.4. The standard automatic adaptive approach to numerical quadrature

The standard AAQ implements the refinement of $\Pi_N[a,b]$ as a *subrange binary tree* the evolution of which is controlled by an associated *priority queue*. The binary tree *initialization* equates the root with the input integration domain $[a,b]$ over which a first global $\{Q_1, E_1 > 0\}$ output is computed. If the termination criterion (2)–(3) is not fulfilled, then a recursive procedure is followed: the priority queue is activated, the resulting root is *bisected*, local estimates $\{q, e > 0\}$ are computed over each resulting sibling, the global quantities $\{Q_N, E_N > 0\}$ are updated, and the end of computation is checked again.

The standard subdivision scheme can be supplemented with a *convergence acceleration algorithm* if the occurrence of an *integrand singularity* was heuristically inferred.

The scrutiny of the existing AAQ implementations into program libraries (IMSL, NAG, SLATEC) or in separate modules (see the evidence reported in [3, 4, 5, 7, 8, 9, 10]) points to the existence of a wide variety of specific integrals (1) which are successfully solved by means of the existing standard AAQ codes.

Taking into account this circumstance, the BAAQ approach to the solution of (1) was upgraded by a decision branch enabling the computation of an enhanced AAQ (EAAQ) solution in cases where the simpler AAQ approach is sufficient to get quick fulfilment of the termination criterion (2). The most important improvements done within EAAQ with respect to the standard AAQ approach described in QUADPACK [3], which substantially increase reliability while keeping on the code fast, are listed below.

2.5. Ways of enhancing reliability of the AAQ codes

(i) *A more reliable local error estimator within Clenshaw-Curtis quadrature.*
Within the local quadrature rules, the assessment of the magnitude of the local quadrature error $e > 0$ uses *probabilistic arguments* and this feature naturally results in the dependence of the reliability of the local error output on the details of the integrand variation inside the current integration range.
Extensive numerical results [8] providing comparisons of three kinds of local error estimators for Clenshaw-Curtis (CC) numerical quadrature (CC–like [11] — see also [12, 13, 14], Sloan-Smith–like [15], and QUADPACK–like [3]) pointed to the existence of significantly different reliability levels of the three kinds of local error estimators. The most reliable one was found to be the CC–like error estimator involving unweighted contributions coming from the four highest order terms of the Chebyshev series expansion of the integrand over the current range. For this reason, our implementation of CC local quadrature rules uses namely this kind of local error estimator.

(ii) *Handling the priority queue.*
The already defined subranges in the partition (4) are divided into two classes: subranges bearing *significant meaning* $\{q, e > 0\}$ at output and subranges bearing *conventional meaning* $\{q = 0, e = \Omega\}$ at output, where Ω denotes a very large positive number close to the machine overflow. Being characterized by the highest possible local error estimate, a subrange carrying conventional meaning is put by the priority queue ordering algorithm at the beginning of the priority queue, such that it is bisected before any subrange bearing significant meaning at output.

(iii) *Conditional activation of the global termination criterion.*
The output brought by a subrange marked as conventional is useless for the update of the global pair $\{Q_N, E_N\}$. Therefore, in this case Q_N and E_N are left to their previous values and the check of the global termination criterion (2) is not attempted. An easy implementation of this feature was obtained by keeping record of the number of subranges carrying conventional meaning and by attempting the check of (2) only provided this number reaches the floor value zero.

(iv) *Stabilization of the local quadrature sums as Riemann sums.*
The assignment of significant or conventional meanings to the local quadrature rule outputs $\{q, e > 0\}$ is easily implemented by means of two criteria checking the stabilization of the local quadrature sums as Riemann sums.
The first criterion is implemented inside the procedures devoted to the computation of the local quadrature rule outputs. It checks, for each sibling generated by the bisection of the current root node of the subrange binary tree, the relationship inside the pair $\{q, e > 0\}$.

The condition

$$|q| > \varphi_{min} \cdot e, \quad \varphi_{min} = 2^{-8}, \tag{5}$$

rejects the computed $\{q, e\}$ values and enforces the assignment of conventional output for the current sibling because the estimated accuracy of the computed local quadrature sum stays below the two accurate significant figure threshold requirement.

The second criterion is enforced inside the global control procedure. It checks the relationship between the output $\{q_p, e_p\}$, carried by the parent subrange which was brought to the root position in the subrange binary tree by the priority queue ordering, and the summed outputs $\{q_d, e_d\}$ coming from its two descendant subranges. If the *variation* of the quadrature sum outputs *exceeds* the summed local errors (where a value $e_p = 0$ is assigned if the parent subrange carries conventional meaning),

$$|q_p - q_d| > e_p + e_d, \tag{6}$$

then conventional meanings are assigned to both descendants.

The fulfilment of one or both criteria entails corresponding modification of the number of subranges recorded as carrying conventional meaning.

3. Bayesian inferences on the classes of integration domain lengths

3.1. Critical issues of the Bayesian automatic adaptive quadrature

While the reliability of the EAAQ code was significantly enhanced in comparison with those of the standard AAQ codes, there remained critical issues which could not be solved within it. Thus, in the overwhelming part of the practical situations, the refinement of the partition (4) by subrange bisection cannot enforce the location of the inner discontinuity points of the integrand at the inner abscissas of $\Pi_N[a, b]$ and to secure in this way fast convergence of the quadrature sums or of the convergence acceleration algorithms. As a consequence, in such a cases, the derivation of outputs of the local quadrature sums such as to meet the termination condition (2) is well-known to be highly ineffective. Even worse, in most cases, the computation ends with failure.

The BAAQ approach tries to overcome this difficulty by resolving the location of the offending abscissas to machine accuracy and including the obtained results into the inner abscissas entering the partition $\Pi_N[a, b]$. We have learned how to solve efficiently such problems [16, 17].

During the development of the corresponding code, we had to learn about the conformity of the hardware and software environment at hand with the IEEE 754 standard (1985 version or 2008 revision) which governs the binary floating-point arithmetic (number formats, basic operations, conversions, exceptional conditions).

Subsequent investigations have unveiled the dependence of the *degree of precision* of the floating-point algorithms on the position and extension of the integration domain on the real number axis. This topics deserves special consideration, since it provides the key to the consistent implementation of BAAQ algorithms over all possible integration domains $[a, b]$.

3.2. Forward floating-point degree of precision of a quadrature sum [18]

The derivation of the local quadrature rule outputs $\{q, e > 0\}$ within the existing AAQ algorithms is done by use of *interpolatory quadrature sums*. An interpolatory quadrature sum of the *algebraic degree of precision* d solves exactly the integrals over the fundamental power set $\{1, x, x^2, \cdots, x^d\}$.

Under computations done over $\mathbf{R}$, the field of real numbers, d is a characteristic invariant of the given interpolatory quadrature sum: the integral over the probe integrand $\pi_d(x) = \sum_{l=0}^{d} x^l$ is solved exactly such that the individual monomial contributions to the output are *distinct from each other* and *non-negligible* irrespective on the *extent* and *localization* of the integration domain $[a, b]$ over the real axis.

However, under floating-point computations on a computer using t-bit length of the significand approximation of the numbers within the continuous field $\mathbf{R}$, the quantity d *ceases to be an invariant* of the interpolatory quadrature sum. Nevertheless, an *integration domain dependent* probe integrand $\pi_{d_{\rm fp}}(x) = \sum_{l=0}^{d_{\rm fp}} x^l$ can be defined for which non-negligeable, distinguishable from each other, individual monomial contributions to the computed output are present.

The formal definition of the *floating-point degree of precision*, $d_{\rm fp} = d_{\rm fp}[a, b]$, was done in [18]. This definition asks for the preservation of the non-negligibility of the individual monomial contributions to $I[a, b]\pi_{d_{\rm fp}}$ at both ends of the power series set $\{1, x, x^2, \cdots, x^{d_{\rm fp}}\}$, $d_{\rm fp} \le d$.

For the implementation of the BAAQ algorithm over all possible scales of the integration domain lengths, the fact that a lowest degree subset of monomials entering $\pi_{d_{\rm fp}}(x)$ might result in negligible contributions to the quadrature sum is irrelevant. The essential feature is the preservation of the non-negligibility of the monomial contributions at the high degree end of the power series set. This results in the following *definition* of the *forward floating-point degree of precision* of an interpolatory quadrature sum:

1^0. Let $I[\alpha, \beta]f$ denote the input integral of interest, defined over a finite integration range $[\alpha, \beta] \subset \mathbf{R}$.

2^0. Let $q[\alpha, \beta]f$ denote an interpolatory quadrature sum of the algebraic degree of precision d, which solves $I[\alpha, \beta]f$ through floating point computations over a set of machine numbers characterized by a t-bit significand.

3^0. Let $fl(a)$ denote the floating point approximation of $a \in \mathbf{R}$, and let

$$X = \max\{fl(|\alpha|), fl(|\beta|)\},\ X > 0, \quad \rho = fl(|\beta - \alpha|/X),\ 0 < \rho \le 2\,. \tag{7}$$

4^0. If $\xi > 0$ stands for either X or ρ, we define

$$d_\xi = \begin{cases} d & \text{iff} \quad \xi \ge x_m \\ \lceil \ln \varepsilon_0 / \ln \xi \rceil & \text{iff} \quad \xi < x_m\,, \end{cases} \tag{8}$$

where $x_m = fl(\varepsilon_0^{1/d})$, $\varepsilon_0 = 2^{-t}$, and $\lceil a \rceil$ is the ceiling of $fl(a)$.

5^0. Then the *floating point degree of precision*, $d_{\rm fp}$, associated to $q[\alpha, \beta]f$ is the positive integer

$$d_{\rm fp} = \min\{d_X, d_\rho\}\,, \tag{9}$$

with d_X and d_ρ computed from Eq. (8) for the terms X and ρ of the pair (7).

3.3. Features of the machine number distributions inside integration domains

The floating-point degree of precision is a straightforward consequence of the floating-point representation of the infinite continuous set of the real numbers by a *finite set of machine numbers*, each of which being characterized by a *fixed length of the significand* and a *finite range of variation of the characteristic* [19, 20].

The distribution of the set of non-negative machine numbers is *discrete* and *highly non-uniform*: denser toward zero and scarcer toward the largest value machine number end. A similar property holds true for the set of non-positive machine numbers.

In the monographs devoted to the AAQ, [3, 4, 5], it is, however, assumed that there is a *continuous, uniform distribution* of the possible values of the argument of the integrand inside the integration domain, such that we can always choose convenient sets of distinct quadrature knots at which the integrand is interpolated by a polynomial for which a quadrature sum is derived.

The need of abandoning the algebraic degree of precision in favour of the floating-point degree of precision is at variance with this assumption. A consistent and economic choice of the interpolatory quadrature sums for BAAQ solution is to make use of the available floating-point degree of precision over the current integration interval.

Over *large enough* integration domains, the distances inbetween adjacent quadrature knots exceed the distances inbetween neighbouring machine numbers by many orders of magnitude such that the AAQ assumption looks justified.

However, if the length of the integration interval falls below some position dependent threshold (and results in a floating-point degree of precision less than two units), then the AAQ assumption obviously fails.

In the next subsection, we show that elementary intervals can be defined inside which the discrete set of inner machine numbers is nearly uniformly distributed.

3.4. Ranges of nearly uniform machine number distributions

For the sake of concreteness, we assume henceforth that the computations are done in *binary64* format as defined by the IEEE 754-2008 standard.

Therefore, the physical length of the significand is 52 bits and, with the addition of the hidden bit, the relative error in the approximation of any real number sufficiently far from zero does not exceed $\bar{\varepsilon}_0 = 2^{-53}$.

The minimum positive normal floating-point number corresponds to the underflow threshold $u_0 = 2^{-1022}$.

Given u_0 and $\bar{\varepsilon}_0$, the ratio $v_0 = u_0/\bar{\varepsilon}_0$ provides a floating-point interval around zero, $(-v_0, v_0)$, within which the normalized floating-point values are evenly spaced.

The relative distance between two neighbouring machine numbers is $\varepsilon_0 = 2\bar{\varepsilon}_0 = 2^{-52}$ (see [1] for the experimental proof of a consistency criterion checking this value).

Given the floating-point number $\eta_0 \notin (-v_0, v_0)$, lateral intervals (η_0^-, η_0) and (η_0, η_0^+) can be defined inside which the *floating-point distance* equates $|\eta_0|\varepsilon_0$.

In order to define how far from η_0 can stay η_0^- and η_0^+, the negligibility of second order effects is to be secured. The relative distance ε_d corresponding to this aim is found from the condition $\varepsilon_d^2 < 2^{-68}$, where $64 + 1 = 65$ denotes the number of the binary bits in the CPU floating-point unit and further three units were added to skip effects coming from the guard digits.

We get $\varepsilon_d = 2^{-34}$, a result which shows that the 18 *least significant binary bits* can have only *linear effects* upon the floating point operations.

4. Integration domain adapted BAAQ algorithms

The results derived in the previous section allow the implementation of a BAAQ approach which selects local quadrature rules adapted to the characteristic feature of each integration interval $[\alpha, \beta] \subseteq [a, b]$.

A number of five distinct cases were identified. They are resolved in a hierarchical order following the description pattern given below.

(1) *Vanishing integration domain length*
The condition $b \neq a$ is checked at the beginning of the computation.
If $b = a$ is found to hold, then a vanishing value of the integral with zero associated error are returned.

(2) *Void open integration interval* (a, b)
If $|b - a| < \max\{|a|, |b|, v_0\} \cdot \varepsilon_0$, then the cardinality of the machine numbers inside (a, b) is zero.

Journal of Physics: Conference Series **627** (2015) 012010 doi:10.1088/1742-6596/627/1/012010

The code returns a warning message together with a result $\{Q, E = |Q|\}$, where Q is computed using the trapeze quadrature sum for the ends of $[a, b]$.

(3) *Integration domain of* microscopic *length*
If $|b - a| < \max\{|a|, |b|, v_0\} \cdot \varepsilon_d$, then there is a (nearly) uniform discrete distribution of machine numbers inside $[a, b]$.
Estimates of both Q and E are derived using composite trapeze quadrature sums.

(4) *Integration domain of* mescoscopic *length*
A value $d_{\text{fp}} < 5$ was found for the forward floating-point degree of precision.
The cardinality of the machine numbers inside (a, b) exceeds $2^{18} \approx 2.6 \times 10^5$.
Composite Simpson, trapeze and centroid quadrature sum outputs are used to check the integrand conditioning and to compute estimates for Q and E.

(5) *Integration domain of* macroscopic *length*
This is the case investigated within the standard AAQ. Most of the BAAQ developments have been described in [1].
The important addition described in the section 2.5 above is the inclusion of the possibility to solve easy integrals within the BAAQ approach.
The principal new BAAQ progress with respect to [1] was the derivation of local error estimates using *the principle of redundancy.*
The QUADPACK approach [3, 5], based on the use of two related to each other quadrature sums (such that the quadrature knots of the lower algebraic degree quadrature sum are a subset of the quadrature knots of the higher algebraic degree quadrature sum) was abandoned. The local error estimate was found from a pair of quadrature sums sharing a single quadrature knot, the centre $\gamma = (\beta + \alpha)/2$ of the current integration range $[\alpha, \beta]$, while the fractionary reduced quadrature knots are statistically independent of each other.
The main quadrature sum output, q_{CC}, was derived by means of CC-33, a closed 33-knot quadrature sum spanned by the set of reduced quadrature knots equating the extremal points of the 32-nd degree Chebyshev polynomial of the first kind over $[-1, 1]$. The resulting interpolatory polynomial is known to be close to the minimax polynomial. As a consequence, q_{CC} provides highly accurate solutions for the integrals involving smooth integrand functions inside the current integration range $[\alpha, \beta]$.
The algebraic degree of precision of CC-33 is $d_{CC} = 32$.
The auxiliary quadrature sum output, q_{GK}, was derived by means of the Gauss-Kronrod GK10-21 of QUADPACK [3], an open 21-knot quadrature sum spanned at reduced Gauss-Kronrod quadrature knots.
The algebraic degree of precision of GK10-21 is $d_{GK} = 31$.
An immediate consequence follows: the accuracy predictions of the local error estimate are significantly higher than those of the conventional QUADPACK error estimates, where they were hindered by the less accurate quadrature sum entering the local error estimator.
After checking out the absence of the exceptional cases mentioned in [1], a local error estimator, equated to the modulus of the difference $|q_{CC} - q_{GK}|$, was used.
If the termination criterion (2) was not fulfilled after the attempt to compute the integral over the whole integration domain $[a, b]$, then the relative error estimate was compared to the conventional accuracy threshold $\tau_a = 2^{-11} \approx 0.49 \cdot 10^{-3}$.
Under disagreement smaller than τ_a, the integral was assumed to be 'easy' and its solution by means of EAAQ was attempted. The Bayesian inference that EAAQ is the code of choice is not definitive. It may be denied under refinement of the partition (4) and the Bayesian approach for solving 'difficult' integrals be activated again.
Under disagreement exceeding τ_a, the BAAQ procedure using the priority queue approach described in [1] was used.

5. Conclusions

The Bayesian inferences reported in this paper significantly extend the coverage of the BAAQ approach to the solution of Riemann integrals (1).

The scrutiny of the possible machine number approximations of abscissa values inside the input integration domain (partially based on the use of the floating-point degree of precision) unveiled the occurrence of five classes of integration domains entering the quadrature problems: zero-length, open void, microscopic, mesoscopic, and macroscopic. The mentioned classes point to different extensions of the integration domains, increasing from the first to the last. The order of class identification cannot be arbitrary: it precisely follows the present order of enumeration. Finally, each class needs specific advancement to the solution, based on the use of different sets of local quadrature sums optimally adapted to the derivation of the output.

In the most complex, macroscopic case, besides the already developed Bayesian solution for 'difficult' integrals (characterized by the occurrence of inner integrand discontinuities) [1], a new decision branch was proposed which allowed the inclusion of 'easy' integrals as well inside a same code.

The reliability of the EAAQ code used for solving 'easy' integrals contains four kinds of improvements which supersede the reliability of the standard AAQ codes [3, 5, 6] and allow the turn back to the full BAAQ code under the detection of a previously unnoticed difficulty.

Acknowledgments

The work was supported within the JINR topic 05-6-1119-2014/2016 and the Romania–LIT Hulubei-Mashcheryakov Programme, JINR Orders 94/17.02.2014, p.25, 95/17.02.2014, pp.76, 77, and 96/17.02.2014, pp. 86–89.

References

[1] Adam Gh and Adam S 2012 Bayesian automatic adaptive quadrature: An overview *Mathematical Modeling and Computational Science LNCS* vol 7125, ed Gh Adam, J Buša and M Hnatič (Heidelberg: Springer) pp 1–16

[2] Adam Gh and Adam S 2009 Principles of the Bayesian automatic adaptive quadrature *Numerical Methods and Programming: Advanced Computing* **10** 391–7 http://num-meth.srcc.msu.ru

[3] Piessens R, deDoncker-Kapenga E, Überhuber C W and Kahaner D K 1983 *QUADPACK, a subroutine package for automatic integration* (Berlin: Springer Verlag)

[4] Davis P J and Rabinowitz P 1984 *Methods of Numerical Integration* Second edition (Orlando (Fla), USA: Academic Press)

[5] Krommer A R and Ueberhuber C W 1998 *Computational Integration* (Philadelphia: SIAM)

[6] *NAG Library Manual* Mark 23 section X02 http://www.nag.co.uk/numeric/FL/nagdoc_fl23/html/FRONTMATTER/manconts.html

[7] Lyness J N 1983 When not to use an automatic quadrature routine *SIAM Rev.* **25** 63–87

[8] Adam Gh and Nobile A 1991 Product integration rules at Clenshaw-Curtis and related points: a robust implementation *IMA J. Numerical Analysis* **11** 271–96

[9] Adam Gh 1993 Case studies in the numerical solution of oscillatory integrals *Rom. Journ. Phys.* **38** 527–38

[10] Adam Gh, Adam S and Plakida N M 2003 Reliability conditions on quadrature algorithms *Comp. Phys. Comm.* **154** 49–64

[11] Clenshaw C W and Curtis A R 1960 A method for numerical integration on an automatic computer *Numer. Math.* **2** 197-205

[12] Filippi S 1964 Angenäherte Tschebyscheff-Approximation einer Stammfunktion-eine Modifikation des Verfahrens von Clenshaw und Curtis, *Numer. Math.* **6** 320–8

[13] Piessens R and Branders M 1975 Algorithm 002. Computation of oscillating integrals *J. Comp. Appl. Math.* **1** 153–64

[14] Hasegawa T, Torii T and Ninomiya I 1983 Generalized Chebyshev interpolation and its application to automatic quadrature *Math. Comp.* **41** 537–53

[15] Sloan I H and Smith W E 1980 Product-integration with the Clenshaw–Curtis points: Implementation and error estimates *Numer. Math.* **34** 387–401

[16] Adam Gh, Adam S, Ţifrea A and Neacşu A 2006 Resolving thin boundary layers in numerical quadrature *Romanian Reports in Physics* **58** 155–66
[17] Adam Gh and Adam S 2008 The boundary layer problem in Bayesian adaptive quadrature *Physics of Particles and Nuclei Letters* **5** 269–73
[18] Adam S and Adam Gh 2012 Floating-point degree of precision in numerical quadrature *Mathematical Modeling and Computational Science LNCS* vol 7125, ed Gh Adam, J Buša and M Hnatič (Heidelberg: Springer) pp 189–94
[19] Goldberg D 1991 What every computer scientist should know about floating-point arithmetic *ACM Computing Surveys* **23** 5–48 http://www.validlab.com/goldberg/paper.pdf
[20] Kahan W 1997 Lecture notes on the status of IEEE standard 754 for binary floating-point arithmetic http://www.eecs.berkeley.edu/ wkahan/ieee754status/IEEE754.PDF

Journal of Physics: Conference Series **627** (2015) 012011 doi:10.1088/1742-6596/627/1/012011

Latest results on the Higgs boson discovery and investigation at the ATLAS-LHC

Nguyễn Anh Kỳ[1,2] **and Nguyễn Thi Hồng Vân**[1]

[1] Institute of physics, Vietnam academy of science and technology, 10 Dao Tan, Ba Dinh, Hanoi, Vietnam

[2] Laboratory for high energy physics and cosmology, Faculty of physics, College of science (VNU university of science), Vietnam national university, 334 Nguyen Trai, Thanh Xuan, Hanoi, Vietnam

E-mail: anhky@iop.vast.ac.vn, nhvan@iop.vast.ac.vn

Abstract. On 04 July 2012, ATLAS and CMS, the two biggest LHC collaborations, announced the discovery of a new boson with a mass around 125 GeV. With the preliminarily delivered information, this particle resembles the Higgs boson (of the standard model) playing a key role in the symmetry breaking mechanism generating particle masses in the standard model. To confirm if the newly discovered particle is really the Higgs boson it is necessary to do further and detailed investigations to precisely determine its nature and properties. Here we will summarize some of the latest results of these investigations. If the new particle is the Higgs boson, the standard model will get the full recognition as a very successful model, otherwise, another model may be needed to replace the standard model in order to explain all the observed phenomena. The results obtained by ATLAS (and CMS) for last two years have almost shown that this new particle can be identified with the Higgs boson.

1. Introduction

The Brout-Englert-Higgs boson of the standard model (SM), or just, the Higgs boson, for short, plays a central role in the symmetry breaking scheme giving masses to particles (in the SM)[1]–[5]. Until recently, however, the Higgs boson had been the only particle in the SM not found yet experimentally. It is why searching for the Higgs boson has been an important mission of experimental particle physics. Searching for the Higgs boson was also one of the main goals of building the LHC (*Large Hadron Collider*), the ever most expensive, most powerful and largest particle accelerator. The LHC has four main collaborations, namely, ATLAS, CMS, LHCb and ALICE among which ATLAS (*A Toroidal LHC Apparatus*) and CMS (*Compact Muon Solenoid*) are the two biggest ones. In the summer of 2012 a breaking scientific news was spreading around the world when the collaborations ATLAS and CMS announced the discovery of a new boson of mass of around 125 GeV which might be (identified with) the Higgs boson of the standard model [6, 7]. If the newly discovered boson is identified with the Higgs boson, the SM becomes a fully confirmed model with all its particle content found experimentally. Since its discovery this particle has been intensively investigated so that its properties can be precisely determined, and, thus, its exact nature can be established. Here we will briefly review some of the latest results on investigating the new boson. After the first observations [6, 7] of the new boson via its decays to gauge boson pairs ($\gamma\gamma$, WW and ZZ), its mass, spin-parity and other properties have been more precisely measured and investigated by different collaborations, ATLAS, CMS and Tevatron. One of the important investigations to confirm if the new particle is really the Higgs boson is to investigate if it also couples to fermions and if these couplings are compatible with the SM in which the couplings between the Higgs boson and the fermions give masses to the latter. Recently, both ATLAS and CMS gave a positive answer of this question but below we will concentrate our discussions on the ATLAS experiment.

With the results obtained so far it can be almost sure that the newly discovered particle is the long sought Higgs boson. It has a mass of about 125 GeV, spin-parity $J^P = 0^+$ and couplings to gauge bosons and fermions as predicted by the SM. The discovery of this particle showing that an elementary scalar particle exists in the Nature, is very meaningful for particle physics as, besides that it generates particle masses, all scalar particles found so far are not elementary but composite. This discovery may have cosmological and other consequences as well but they are beyond the scope of our report. Before going to more physical discussions in section 3, let us make in the next section a technical overview on the LHC and the ATLAS detector.

2. Brief information of the LHC and the ATLAS

Here we will present a brief introduction to the LHC and the ATLAS detector. More information about these facilities can be found from their official websites [8, 9].

The LHC, located at CERN (*European organization for nuclear research*), the world's leading laboratory for particle physics, is the biggest (in size, cost and collision energy) particle accelerator ever built by human. It is installed in a tunnel of 27 km long and at a depth of 50 – 100 m under the French-Swiss border near Geneva. It was designed to accelerate and collide two proton beams at the maximal center-of-mass energy of 14 TeV ($\sqrt{s} = 14$ TeV). The collision energy reached in the first run (2009 – 2012) of the LHC was 8 TeV ($\sqrt{s} = 8$ TeV). The designed maximal energy (14 TeV) is expected to be reached in the second run which will start in 2015. Besides the record particle collision energies, the LHC has achieved many technological records such as a superstrong magnetic field (8.4 T, or about 200 000 times stronger than the Earth's magnetic field), the highest vacuum ($10^{-10} - 10^{-11}$ mbar, in the order of that on the Moon surface), the lowest temperature (1.9 K, or lower than that in the outer interplanetary space, 2.7 K), the highest temperature (5.5 trillion degrees Celsius, or near 350 000 times higher than the temperature in the center of the Sun), etc.

The LHC has four main detectors among which ATLAS and CMS are the two biggest and general-purpose ones (see Fig. 1 for a scheme of how the detectors are located on the LHC ring). The ATLAS and CMS detectors are based on the same operation principle and have similar general structures and purposes. The detailed differences are their sizes, masses, magnetic fields used, sensitivities of sub-detectors, etc.. These detectors are huge and complex high-technological facilities, for example, the ATLAS detector has four sub-detectors which in turn consist of many layers with a total mass of about 7000 tones and an overall size of about 25m (diameter) × 46m (length), while the CMS detector is heavier (13000 tones) but smaller (15m × 22m), thus, the name *"compact"*. Because of materials used, the ATLAS detector, compared with the CMS detector, has a more sensitive hadron calorimeter (thus, a better jet resolution) but a less sensitive electromagnetic calorimeter (thus, a worse e/γ resolution). The ATLAS inner detector is surrounded by a 2T magnetic field, while the CMS one - by 4 T magnetic field. That means the CMS inner detector has a better momentum resolution but restricts the design of

Journal of Physics: Conference Series **627** (2015) 012011 doi:10.1088/1742-6596/627/1/012011

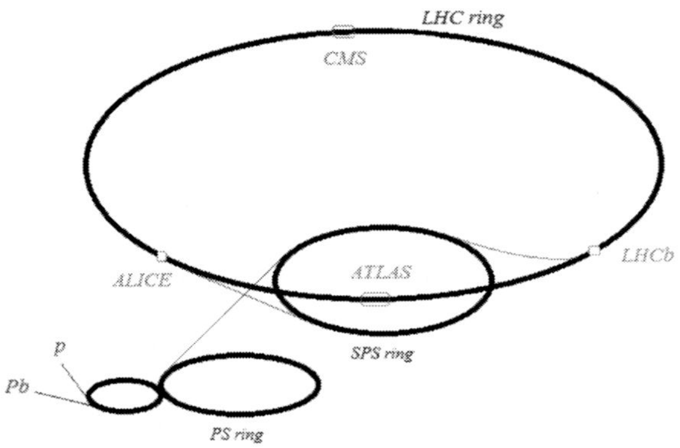

Figure 1: LHC ring and detectors .

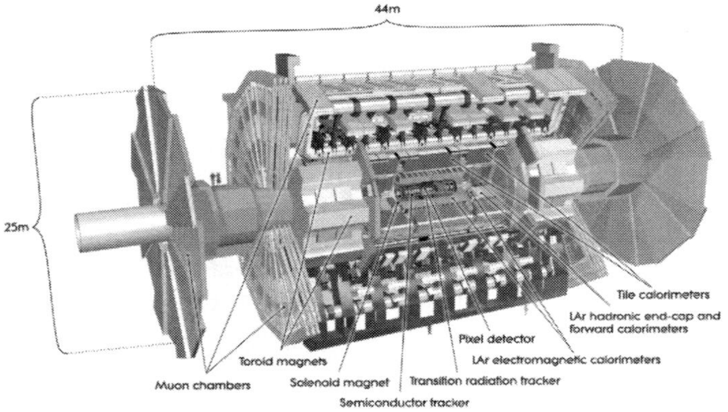

Figure 2: ATLAS detector layout [9].

other components. The research scopes of the ATLAS and the CMS spread on the test of the SM and searching for new physics (beyond the SM): precision measurements of particle parameters and properties, search for the Higgs boson, extra dimensions, supersymmetry, dark matter, etc. Each of these collaborations is participated by more than 3000 scientists and engineers from more than 170 institutions in about 40 countries. Here, as an example, we will make a description of the ATLAS detector. More information of the ATLAS and the CMS detectors and a detailed comparison between them can be got from [10].

As an LHC detector, the ATLAS detector is a complex research facility containing many parts with different functions (see its layout in Fig. 2). This detector has four main components which are ever-larger concentric cylinders, namely, the inner detector (ID), the calorimeters (CM's), the muon spectrometer (MSM) and the magnet systems (MS), surrounding the proton beam axis. Without going to physical and technical details, let us recall general structures and functions of these components.

The ID consisting of three high-resolution parts (the pixel detector, the semi-conductor tracker and the transition radiation tracker (TRT)) used to track and identify charged particles, is the innermost component of the ATLAS detector, surrounded by a solenoidal superconducting magnet system and surrounding the interacting point at the centre where collisions of proton beams take place. The ID measures positions and momenta of charged particles at the pseudo-rapidity range $|\eta| < 2.5$ (in which the TRT covers the range $|\eta| < 2.0$).

The function of the CM's surrounding the ID is to measure energies of (easily stopped) particles by absorbing these energies. This component of the ATLAS detector is divided into two parts: the electromagnetic calorimeter (EC) and the hadronic calorimeter (HC). The EC measures with high precision energies and locations (including trajectories) of particles sensitive to electromagnetic interaction such as photons and charged particles. The HC is designed to measure the energies of those particles sensitive to the strong interaction. It has no high precision as the EC but it can measure the particles at the range $|\eta| < 4.9$, primarily hadrons, which the EC cannot catch.

The MSM is the outermost layer of the ATLAS detector. It is a very large system surrounding the CM's and tracking outgoing muons which are the only detectable particles not stopped by the CM's. The MSM measures the paths of muons and their momenta with a very high precision. It is composed from three parts: a set of large superconducting toroidal magnets, a set of chambers tracking with high spatial precision the outgoing muons, and a set of chambers triggering particles with high time-resolution. The MSM tracks muons at the range $|\eta| < 2.4$ (triggering chambers) and $|\eta| < 2.7$ (tracking chambers).

The MS around the ID and the MSM are used to produce appropriate magnetic fields to bend trajectories of (charged) particles so that their momenta and charges can be determined. The solenoidal magnets, surrounding the ID, can produce 2 Tesla magnetic fields with a peak at 2.6 T, while the magnetic fields produced by the toroidal magnets around the MSM are 0.5 T (by the barrel coils) and 1 T (by the end-cap coils).

For a summary and a general imagination, a simplified view on the structure of the ATLAS detector and how a particle can be detected in the ATLAS detector is shown in Fig. 3 (see more in [8, 9, 10]).

With its very large and extremely fine structure, the ATLAS (along with the CMS) could detect for the first time a new boson which now, after a number of more precise investigations of its masses, spin-parity (J^P) and other production and decay information, can be almost identified with the long-sought standard model Higgs boson. It is a scalar particle having a mass of about 125 GeV, $J^P = 0^+$ and coupling to gauge bosons and fermions (quarks and leptons) as expected by the SM. These investigations have been done on the base of the analysis of the data collected in 2011 (for $\sqrt{s} = 7$ keV) and 2012 ($\sqrt{s} = 8$ keV). Now, with high confidence and for convenience we can call hereafter the new boson the Higgs boson.

3. Searching for the Higgs boson: Why and How ?

As mentioned above, one of the missions of ATLAS and CMS is to search for the Higgs boson which can be treated as the heart of the symmetry breaking mechanism giving masses to particles. It is very important because, for example, the fact that the gauge bosons (W and Z) get masses makes the weak interaction short-ranged, otherwise, the structures like atoms, thus the Universe, could not be formed, and many processes in the Nature such as the reactions

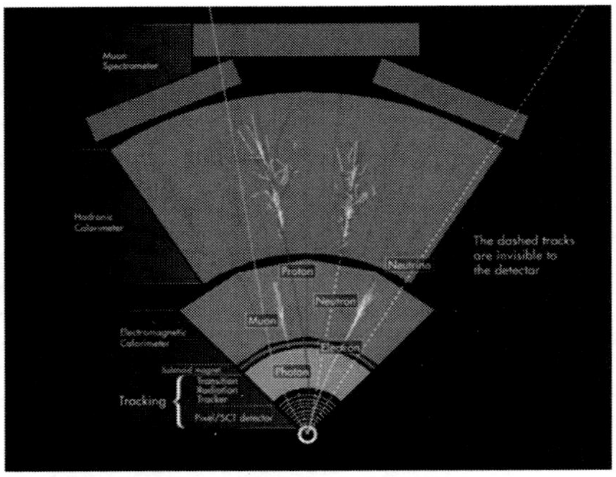

Figure 3: General structure of the ATLAS detector and particle traces [9].

on the Sun, making, in particular, the life on the Earth possible, could not occur. If the Higgs boson does not exist one must work out another symmetry breaking scheme or to deal with another mechanism to generate masses of particles (there have been such attempts but we do not discuss them here). The new boson is also the first real example of an existing fundamental scalar and it is believed to play an important role in particle physics and cosmology. Therefore, the existence or the nonexistence of the Higgs boson may decide the fate of the SM and other theories and models. All that explains why the Higgs boson has been one of the most sought particles for nearly 50 years and its discovery could be classified by some people as one of the most remarkable and important scientific discoveries in the last 100 years. Until the discovery of the Higgs boson the belief in its existence has increased over time as more and more predictions of the SM have been experimentally confirmed. This belief gave a strong motivation for searching for the Higgs boson and, therefore, for building the LHC.

To identify a particle it is important to determine all its basic characteristics including its mass which may be a priori estimated by a theory or constrained by some conditions. For the Higgs boson (H) until its discovery, different theoretical constraints and experimental results (precision measurements) had established several possible ranges of its mass (m_H). In the framework of this short report we do not discuss (because of a limited space) Higgs-type bosons in physics beyond the SM as our topic here is to see if the new boson discovered two years ago by ATLAS and CMS is the Higgs boson of the SM.

Theoretically, the Higgs boson mass, or simply, the Higgs mass, which is one of the fundamental parameters of the SM to be determined (experimentally), cannot be directly predicted by the SM but it can be constrained by, for example, the known SM parameters including masses of other particles such as the top quark and the gauge boson W (or Z). The validity of the SM up to the Planck scale requires the Higgs mass $m_H \leq 180$ GeV (triviality bound) [11], while the unitarity constraints [12, 13, 14] put an upper bound of the Higgs mass at around 1 TeV. The Higgs mass is also bounded from below at about 130 GeV (stability bound) [15] when the stability of the Higgs potential is required. If a metastable electroweak vacuum is allowed the lower bound may become smaller, at about 115 GeV [16]. Taking all said above into account, the Higgs mass 125 GeV is far from the triviality bound (that means there is no need of beyond SM physics until the Planck scale) but on the edge of the vacuum stability-instability.

Journal of Physics: Conference Series **627** (2015) 012011 doi:10.1088/1742-6596/627/1/012011

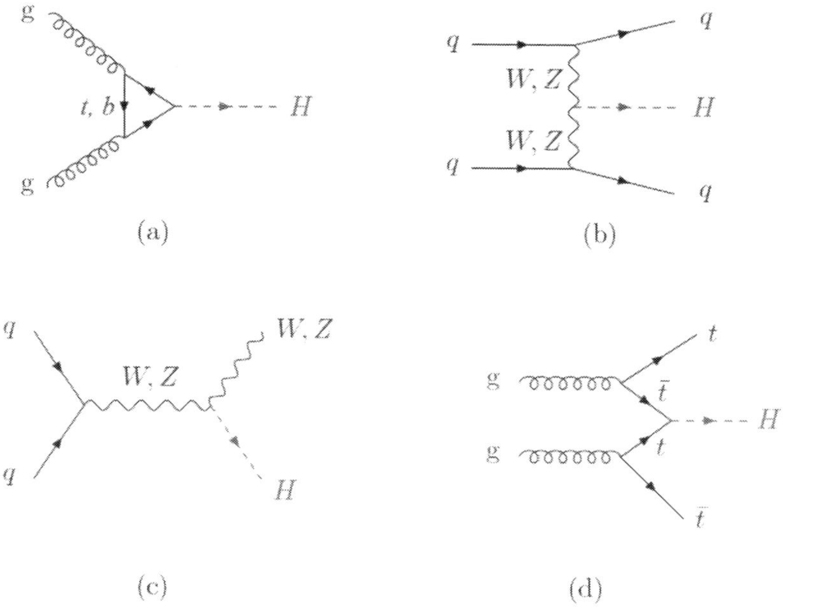

Figure 4: Higgs production diagrams: (a) *gluon fusion*, (b) *W/Z fusion*, (c) *Higgs strahlung*, (d) *top fusion*.

Experimentally, the Higgs mass can be determined indirectly by precision measurements, or "measurements" for short, of electroweak parameters such as the masses of gauge bosons, the top quark mass, Fermi constant, etc., or directly via reconstructing the Higgs mass from its decays. Different collaborations have made measurements to estimate the limits of the Higgs mass which in general have been so far consistent with each other and with the SM. Here we will briefly recall only some results before 4 July 2012 when the Higgs boson's discovery was announced (see more precise information in, for example, [17]). The measurements establish or exclude ranges for the mass of a possible Higgs boson. The lower bound of the Higgs mass established after LEP2 was at about 114.4 GeV [18], while its upper bound given by measurements at LEP, Tevatron and SLC was 152 GeV [19]. The D∅ and CDF collaborations at the Tevatron derived the possible Higgs mass range 115 GeV $< m_H <$ 140 GeV [20]. The combined results [21] from LEP, D∅ and CDF put the possible Higgs mass between the range 115 GeV $< m_H <$ 135 GeV. Results obtained shortly before 04 July 2012 by the ATLAS and the CMS gave a Higgs mass range around 115 GeV $< m_H <$ 130 GeV [22, 23]. The measurements done have given a more and more narrow range of a possible Higgs mass until reaching the final value around 125 GeV which is slightly below the stability bound.

At the LHC the Higgs boson could emerge from a gluon fusion or a weak-gauge-boson (W/Z) fusion or as a production associated with a gauge boson (*Higgs strahlung*) or with top quarks (*top fusion*) among which the gluon fusion followed by the gauge-boson fusion, is the dominant process [17, 24, 25, 26] (see Fig. 4 for Feynman diagrams of these processes). The Higgs boson with a very short life time (of the order $10^{-22}s$ for a mass around 125 GeV) will decay imme-

H decay channel	Branching ratio	Relative uncertainty (%)	Mass resolution (%)
$H \to \gamma\gamma$	2.28×10^{-3}	$^{+5.0}_{-4.9}$	1-2
$H \to ZZ\ (\to 4l)$	2.64×10^{-2}	$^{+4.3}_{-4.1}$	(1-2)
$H \to WW\ (\to l\nu l\nu)$	2.15×10^{-1}	$^{+4.3}_{-4.2}$	(20)
$H \to \tau\tau$	6.32×10^{-2}	$^{+5.7}_{-5.7}$	15
$H \to bb$	5.77×10^{-1}	$^{+3.2}_{-3.3}$	10

Table 1: Sensitive Higgs decay channels at the LHC for $m_H = 125$ GeV [17].

diately into lighter particles. According to the SM its possible decays could include [1] $H \to \gamma\gamma$, $H \to ZZ \to 4l$, $H \to WW \to l\nu l\nu$, $H \to \tau\tau$, $H \to bb$, etc., with branching ratios (BR's), relative uncertainties and mass resolutions given in Table 1 for the Higgs mass $m_H = 125$ GeV [17, 24, 25, 26]. A choice of an optimal channel for the Higgs boson search depends on its sensitivity which in turn depends on several factors such as the cross section of the Higgs production, the branching ratio of the Higgs decay, the resolution of the reconstructed mass, the selection efficiency and the signal-to-background ratio (S/B). All these factors strongly depend on a Higgs mass (or mass range) chosen. Thus, for a given Higgs mass (or mass range), there is/are some channels more preferable for the Higgs search.

The Higgs mass measurements by ATLAS and CMS have been done on the base of the invariant mass reconstruction from the decay channels $H \to \gamma\gamma$ and $H \to ZZ \to 4l$, where $l = e$ or μ. The channels $H \to \gamma\gamma$ and $H \to ZZ \to 4l$ have no large cross sections but they are preferred because their reconstructed final states have high mass resolutions (1-2%) and clean signals (due to the construction of the electromagnetic calorimeter and the muon spectrometer). The use of the other three channels is not excluded but it has the following difficulties. The channel $H \to WW \to l\nu l\nu$ has a relatively large branching ratio but the Higgs mass resolution is very low (20%) because of neutrinos present in the final states (see Table 1). The channels $H \to \tau\tau$ and $H \to bb$, because of a low mass resolution (15% and 10%, resp.) and large backgrounds, have no clean signals. These channels, however, are important in investigating the couplings of the Higgs boson to fermions (see below).

For every measurement, a key problem is to separate the true signal events from the fake, or the background, ones. It can be solved by an estimation of the expected background composition

[1] Hereafter, for short, formal notations like $H \to ZZ$, $H \to WW$, $H \to \tau\tau$ and $H \to bb$ are used but they could be replaced by $H \to ZZ^*$, $H \to W^+W^-$, $H \to \tau^+\tau^-$ and $H \to b\bar{b}$, resp.

and yield in the process measured via a Monte Carlo simulation normalized to the SM theoretical predictions (usually for electroweak-related processes) or by using data (usually for QCD-related processes). A background can be irreducible or reducible (see, for example, [6, 27], for more details). The first type of backgrounds consists of those events with the same final states as those of the signals, while the second one consists of those events with final states misunderstood as the true ones of the signals. As said above, the decays $H \to \gamma\gamma$ and $H \to ZZ \to 4l$ are the "golden" channels for hunting the Higgs boson [6, 7]. Here we briefly count the background (see Fig. 5 and Fig. 6 for a quantitative imagination) in these two Higgs decay channels (while for other channels, see, for example, [27]).

For the channel $H \to \gamma\gamma$, the irreducible background represents the genuine photon pairs produced in Born- ($qq \to \gamma\gamma$), box- ($gg \to \gamma\gamma$) and quark bremsstrahlung ($qg \to q\gamma \to \gamma\gamma$, $gg \to jj\gamma\gamma$) processes, where j denotes a jet. The reducible background for this channel represents γ-jet and jet-jet events in which one or two jets are misidentified as photons, or appear in the decay $Z \to ee$ where the electrons are misidentified as photons. For the channel $H \to ZZ \to 4l$, the irreducible backgrounds come from ZZ^* and $Z\gamma^*$ continuum productions including those in which one of the Z decays into a pair of τ leptons which subsequently decay into lighter leptons. The reducible backgrounds for this channel consist of $4l$ productions from tt and Z+jets (mainly Zbb for the final states $ll + \mu\mu$). More detailed description of the backgrounds for these channels and their estimation can be found in [6] for ATLAS (and similarly, for CMS [7]).

Using the data collected in 2011 and 2012 from proton-proton collisions at the center-of-mass energy $\sqrt{s} = 7$ TeV and $\sqrt{s} = 8$ TeV with an integrated luminosity of 4.5 fb^{-1} and 20.3 fb^{-1}, respectively, ATLAS found the following values of the Higgs mass (see Fig. 5 and Fig. 6):

$$m_H = 125.98 \pm 0.42(\text{stat.}) \pm 0.28(\text{sys.})\ \text{GeV} = 125.98 \pm 0.50\ \text{GeV}, \tag{1}$$

for the channel $H \to \gamma\gamma$ [28, 29] and

$$m_H = 124.51 \pm 0.52(\text{stat.}) \pm 0.06(\text{sys.})\ \text{GeV} = 124.51 \pm 0.52\ \text{GeV}, \tag{2}$$

for the channel $H \to ZZ \to 4l$ [28, 30]. The difference between the mass measurements at the two channels has about 2σ significance or about 4.8% probability. The combined result [28] obtained at $\sqrt{s} = 7$ TeV and $\sqrt{s} = 8$ TeV with an integrated luminosity of 25 fb^{-1} is

$$m_H = 125.36 \pm 0.37(\text{stat.}) \pm 0.18(\text{sys.})\ \text{GeV} = 125.36 \pm 0.41\ \text{GeV}. \tag{3}$$

For the signal strengths, the measurements by ATLAS for the above discussed Higgs decays are in good agreement with the SM: $\mu = 1.17 \pm 0.27$ (for $H \to \gamma\gamma$) [29], $\mu = 1.44^{+0.40}_{-0.33}$ (for $H \to ZZ$) [30] and $\mu = 1.08^{+0.22}_{-0.20}$ (for $H \to WW$) [31]. It is expected to have more precise measurements in the coming time [31], specially, after the start of the second run of the LHC in the beginning of 2015 [32].

The next step in comparing the new boson with the Higgs boson is to check if it also couples to fermions, as predicted by the SM. If the coupling Htt of the Higgs boson to the top quarks can be studied via the top fusion mentioned above, the couplings of the Higgs boson to other, specially, down-type, fermions can be studied via the Higgs decays into these fermions. The ATLAS (also CMS) has made this study on the potential decay channels $H \to \tau\tau$ and $H \to bb$ for $\sqrt{s} = 7$ TeV (4.5 fb^{-1} and 4.7 fb^{-1}, resp.) and $\sqrt{s} = 8$ TeV (20.3 fb^{-1}) [33, 34]. Other its decay channels into fermions are either low ranked because of small BR's (e.g., $H \to \mu\mu$) or kinematically impossible ($H \to tt$). For the channel $H \to bb$ and $m_H = 125.36$ GeV, it

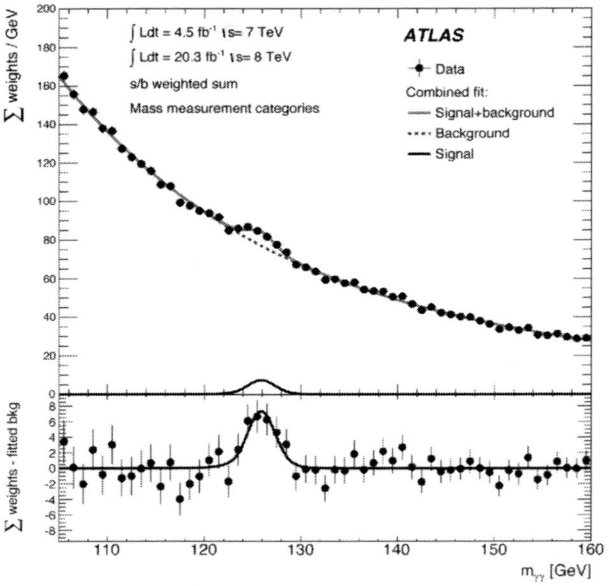

Figure 5: An invariant mass ($m_{\gamma\gamma}$) spectrum in decay $H \to \gamma\gamma$ for the combined $\sqrt{s} = 7$ TeV and $\sqrt{s} = 8$ TeV data and the mass range 105 – 160 GeV [28, 29].

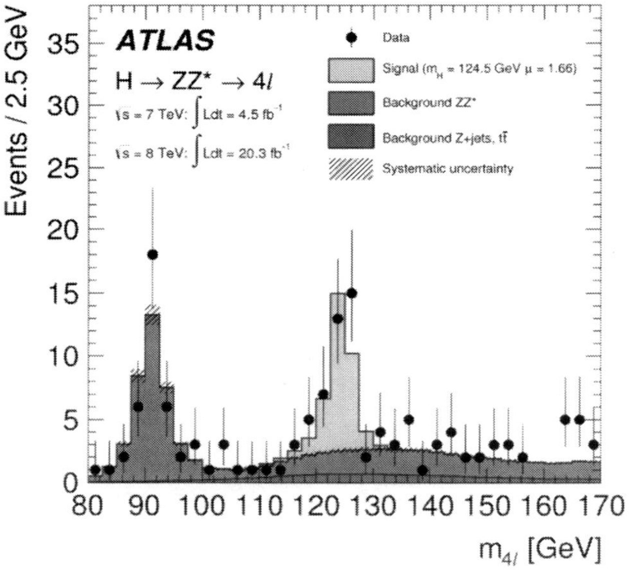

Figure 6: An invariant mass (m_{4l}) distribution in decay $H \to ZZ^* \to 4l$ for the combined $\sqrt{s} = 7$ TeV and $\sqrt{s} = 8$ TeV data and the mass range 80 – 170 GeV [28, 30].

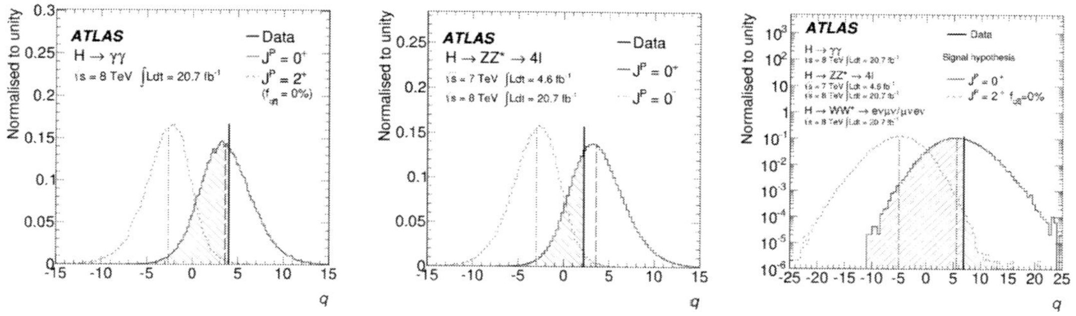

Figure 7: Examples for expected distributions of the logarithm of the ratio of profiled likelihoods (q), under the $J^P = 0^+$ hypothesis in comparison with other spin-parity hypotheses (the observed values are shown by the vertical solid lines) [35, 36].

has been found that the observed deviation from the background is only 1.4σ (compared with the expected 2.6σ) and the signal strength is $\mu = 0.52 \pm 0.32(\text{stat.}) \pm 0.24(\text{syst.})$ [33]. For the channel $H \to \tau\tau$ the results are more convincing as the observed significance of an excess of events over the background is 4.5σ (compared with the expected 3.5σ) and the signal strength is $\mu = 1.42^{+0:44}_{-0.38}$ [34]. The combined results $\mu = 1.30 \pm 0.12 = 0.10 \pm 0.009$ for all channels is in quite good and in agreement with those from CMS [32].

To check if the newly discovered particle is really the Higgs boson it is also necessary to investigate its spin and parity (spin-parity, J^P). First of all, decaying into a neutral pairs of gauge bosons, e.g., two γ's, this particle must be a neutral boson with a spin different from 1. It remains to see if it has a spin-0 or spin-2. This problem has been investigated by ATLAS via the decay channels $H \to \gamma\gamma$, $H \to ZZ \to 4l$ and $H \to WW \to l\nu l\nu$ at $\sqrt{s} = 8$ TeV (20.7 fb^{-1}) and the channel $H \to ZZ \to 4l$ at $\sqrt{s} = 7$ TeV (4.6 fb^{-1}). These investigations (see, for example, Fig. 7) have shown a strong evidence for the scalar (spin-0 and positive-parity) nature, $J^P = 0^+$, of the new boson, as expected for the Higgs boson [35, 36].

Conclusions

We have made a brief review on the progress of searching for the Higgs boson and on the main achievements of the ATLAS collaboration in investigating the boson they discovered with the CMS collaboration over two years ago. It has been shown that the newly discovered boson has a mass about 125 GeV and a spin-parity 0^+. Couplings of the new boson to other bosons and fermions have been also checked in high precision and consistence with the standard model (in combination with the results from CMS). These couplings are very important as they give rise to particle masses. The results obtained allow the new boson to be almost identified with the Higgs boson of the standard model. More discussions on spontaneous symmetry breaking and searching for the Higgs boson can be found in, for example, [37] and references therein.

The CMS collaboration has done similar investigations and has obtained results consistent with those by the ATLAS collaboration but in the framework of this short report we constrain ourselves mainly to the ATLAS investigations. By the same reason, we cannot either discuss physics beyond the standard model in which there may be no or more than one Higgs fields, thus, Higgs bosons, with different masses and properties.

Acknowledgments

The authors would like to thank Phi Quang Van for useful discussions. This work is supported by Vietnam's National Foundation for Science and Technology Development (NAFOSTED) under the grant No 103.03-2012.49.

References

[1] Englert F and Brout R 1964 *Phys. Rev. Lett.* **13** 321.
[2] Higgs P W 1964 *Phys. Lett.* **12** 132.
[3] Higgs P W 1964 *Phys. Rev. Lett.* **13** 508.
[4] Guralnik G S, Hagen C R and Kibble T W B 1964 *Phys. Rev. Lett.* **13** 585.
[5] Higgs P W 1966 *Phys. Rev.* **145** 1156.
[6] Aad G *et al.* (ATLAS collaboration) 2012 *Phys. Lett.* **B716** 1 (arXiv:1207.7214 [hep-ex]).
[7] Chatrchyan S *et al.* (CMS collaboration) 2012 *Phys. Lett.* **B716** 30 (arXiv:1207.7235 [hep-ex]).
[8] The LHC website: http://home.web.cern.ch/topics/large-hadron-collider/.
[9] The ATLAS website: http://www.atlas.ch/.
[10] Froidevaux D and Sphicas P 2006 *Annu. Rev. Nucl. Part. Sci.* **56** 375.
[11] Ellis J, Espinosa J R, Giudice G F, Hoecker A, and Riotto A 2009 *Phys. Lett.*, **B679** 369.
[12] Cornwall J M, Levin D N, and Tiktopoulos G 1973 *Phys. Rev. Lett.* **30** 1286; Cornwall J M, Levin D N, and Tiktopoulos G 1974 *Phys. Rev.* D **10** 1145.
[13] Llewellyn Smith C H 1973 *Phys. Lett.* **B46** 233.
[14] Lee B W, Quigg C, and Thacker H B 1977 *Phys. Rev.* D **16** 1519.
[15] Alekhin S, Djouadi A and Moch S 2012 *Phys. Lett.* **B716** 214.
[16] Espinosa J R and Quiros M 1995 *Phys. Lett.* **B353** 257.
[17] Olive K A *et al.* (Particle data group) 2014 *Chin. Phys.* C **38** 090001 (URL: http://pdg.lbl.gov).
[18] Barate R *et al.* (LEP working group for Higgs boson searches and ALEPH, DELPHI, L3 and OPAL collaborations) 2003 *Phys. Lett.* **B565** 61 (hep-ex/0306033).
[19] LEP electroweak working group (status of March 2012) 2012 http://lepewwg.web.cern.ch/LEPEWWG/.
[20] The TEVNPH working group (for the DØ and CDF collaborations) 2012 *"Updated combination of CDF and D0 searches for standard model Higgs boson production with up to 10.0 fb^{-1} of data"*, FERMILAB-CONF-12-318-E; CDF note 10884; DØ note 6348 (arXiv:1207.0449 [hep-ex]).
[21] Baak M, Goebel M, Haller J, Hoecker A, Ludwig D, Moenig K, Schott M and Stelzer J 2012 *Eur. Phys. J.* C **72** 2003 (arXiv:1107.0975 [hep-ph]).
[22] ATLAS collaboration 2011 *"ATLAS experiment presents latest Higgs search status"* (13 Dec. 2011 seminar), http://www.atlas.ch/news/2011/status-report-dec-2011.html .
[23] CMS collaboration 2011 *"CMS search for the standard model Higgs boson in LHC data from 2010 and 2011"* (13 Dec. 2011 seminar), http://cms.web.cern.ch/news/cms-search-standard-model-higgs-boson-lhc-data-2010-and-2011.
[24] Baglio J and Djouadi A 2011 *JHEP* **1103** 055 (arXiv:1012.0530 [hep-ph]).
[25] Dittmaier S *et al.* (LHC Higgs cross section working group collaboration) 2011 *"Handbook of LHC Higgs cross sections: 1. Inclusive observables"*, arXiv:1101.0593 [hep-ph].
[26] Dittmaier S *et al.* (LHC Higgs cross section working group collaboration) 2012 *"Handbook of LHC Higgs cross sections: 2. Differential distributions"*, arXiv:1201.3084 [hep-ph].
[27] Nguyen Thi Hong Van 2007 *"Discovering the standard model Higgs at the LHC"*, Diploma dissertation, ICTP, Trieste.
[28] Aad G *et al.* (ATLAS collaboration) 2014 *Phys. Rev.* D **90**, 052004 (arXiv:1406.3827 [hep-ex]).
[29] Aad G *et al.* (ATLAS collaboration) 2014 *"Measurement of Higgs boson production in the diphoton decay channel in pp collisions at center-of-mass energies of 7 and 8 TeV with the ATLAS detector"*, CERN-PH-EP-2014-198 (arXiv:1408.7084 [hep-ex]).
[30] Aad G *et al.* (ATLAS collaboration) 2014 *"Measurements of Higgs boson production and couplings in the four-lepton channel in pp collisions at center-of-mass energies of 7 and 8 TeV with the ATLAS detector"*, CERN-PH-EP-2014-170 (arXiv:1408.5191 [hep-ex]).
[31] ATLAS Collaboration 2014 *"Observation and measurement of Higgs boson decays to WW^* with ATLAS at the LHC"*, ATLAS-CONF-2014-060.
[32] J. Ellis, *"The beautiful physics of LHC Run 2"*, arXiv:1412.2666 [hep-ph].
[33] Aad G *et al.* (ATLAS collaboration) 2014 *"Search for the $b\bar{b}$ decay of the standard model Higgs boson in associated (W/Z)H production with the ATLAS detector"*, ATLAS-HIGG-2013-23 (arXiv:1409.6212).

[34] ATLAS collaboration 2014 *"Evidence for Higgs boson Yukawa couplings in the $H \to \tau\tau$ decay mode with the ATLAS detector"*, ATLAS-CONF-2014-061.
[35] ATLAS collaboration 2013 *"Study of the spin of the new boson with up to 25 fb^{-1} of ATLAS data"*, ATLAS-CONF-2013-040, April 2013.
[36] Aad G *et al.* (ATLAS collaboration) 2013 *Phys. Lett.* **B726** 120 (arXiv:1307.1432 [hep-ex]).
[37] Nguyen Anh Ky and Nguyen Thi Hong Van 2015 *"Was the Higgs boson discovered"*, to appear in *Commun. Phys.* (arXiv:1503.08630 [hep-ph]).

Journal of Physics: Conference Series **627** (2015) 012012 doi:10.1088/1742-6596/627/1/012012

Echoes from the past

Jinn-Ouk Gong
Asia Pacific Center for Theoretical Physics, Pohang 790-784, Korea
E-mail: jinn-ouk.gong@apctp.org

Abstract.
Primordial inflation is regarded as the leading paradigm for the very early universe. After quickly reviewing some basic elements, I discuss the current status of inflationary theories especially in the light of effective theory approach and primordial gravitational waves. I also comment on future probes for the early universe.

1. Introduction

Currently, inflation [1] is regarded as the leading candidate to describe the very early universe before the onset of the standard hot big bang evolution. While the current observable universe requires a number of extremely finely tuned initial conditions as suggested by the cosmic microwave background (CMB), inflation can provide these conditions naturally. For example, the so-called horizon problem states that at the moment of the generation of the CMB the observable universe could not be in causal communication but there were $10^4 - 10^5$ causally disconnected regions, so that the chance of having the same temperature with the accuracy of 10^{-5} as the current observations on the CMB is extremely unlikely. According the the inflationary picture, the whole observable universe originated from a single causally connected patch and it is no surprise that the temperature of the CMB is homogeneous. Note that we need then a certain amount of expansion during inflation. This is quantified by the number of e-folds N, which is the logarithmic ratio of the scale factor $a(t)$ at the end of inflation to the initial moment, $N = \log(a_e/a_i)$. Taking into account our ignorance on certain aspects of the early universe, we require $N \gtrsim 50 - 60$.

Inflation does not only provide the necessary initial conditions for the hot big bang evolution of the universe. During inflation, quantum mechanical uncertainties are amplified following the expansion of the universe and they become the classical perturbations in the amount of expansion once the modes of our interest exit the horizon. After inflation and these modes come inside the horizon again, and then experience causal evolution such as gravitational instability. These small inhomogeneities later become the temperature anisotropy of the CMB, the seed of galaxies, and so on – all the observable structure in the universe. Thus, according to inflation, we can probe quantum mechanical signatures relevant for the speculative physics in the early universe by accurate observations on cosmic scales [2].

The perturbations are usually described by the metric fluctuations. Writing the spatial component of the flat Friedmann-Robertson-Walker metric with most general and physical fluctuations,

$$dl^2 = a^2(t)\left\{[1 + 2\mathcal{R}(t, \boldsymbol{x})]\,\delta_{ij} + h_{ij}(t, \boldsymbol{x}) + \cdots\right\} dx^i dx^j \,, \tag{1}$$

where h_{ij} is pure tensor so that it is transverse and traceless. $\mathcal{R}(t, \boldsymbol{x})$ is proportional to the background metric δ_{ij} and is thus scalar in nature. Because it describes the perturbation in the spatial hypersurface, it is also called the curvature perturbation. It is related to the isotropic scaling of space and thus relevant to the temperature fluctuation $\delta T/T_0$ of the CMB. Meanwhile, $h_{ij}(t, \boldsymbol{x})$ describes area-conserving anisotropic stretching of space as can be inferred by its transverse and traceless nature, and possesses two physical degrees of freedom. Thus it is identified as the primordial gravitational waves, leading to quadruple anisotropy in the temperature fluctuation of the CMB and is thus relevant to the polarization of the CMB.

The properties of these perturbations are described by the correlation functions. The inflationary predictions are such that their power spectra are nearly scale invariant with respect to k and almost perfectly Gaussian. Indeed, the running of the power spectrum of the curvature perturbation is described by the spectral index $n_{\mathcal{R}}$,

$$P_{\mathcal{R}}(k) = \left\langle |\mathcal{R}(k)|^2 \right\rangle \propto k^{n_{\mathcal{R}}-4} \tag{2}$$

with k^{-3} factor being just dimensional. Recent observation by Planck constrains $n_{\mathcal{R}} = 0.960 \pm 0.007$ at 1σ level [3]. The departure from Gaussian nature, non-Gaussianity, is frequently parametrized by the so-called non-linear parameter f_{NL}. In the squeezed configuration of the 3-point correlation function, viz. bispectrum of the curvature perturbation, it is given by

$$f_{\text{NL}} = \frac{5}{12} \lim_{k_3 \to 0} \frac{B_{\mathcal{R}}(k_1, k_2, k_3)}{P_{\mathcal{R}}(k_1) P_{\mathcal{R}}(k_3)} = \frac{5}{12}(1 - n_{\mathcal{R}}) , \tag{3}$$

so that f_{NL} is vanishingly small. Indeed, the current bound on f_{NL} is $f_{\text{NL}} = 2.7 \pm 5.8$ at 1σ level, consistent with 0 [4]. For gravitational waves, likewise we can anticipate the power-law form of the power spectrum,

$$P_{\mathcal{T}}(k) = \sum_{s=+,\times} \left\langle |h_s(k)|^2 \right\rangle \propto k^{n_{\mathcal{T}}-3} , \tag{4}$$

where s runs over the two independent polarization states denoted by $+$ and $\times$. An important quantity of interest is the tensor-to-scalar ratio

$$r = \frac{P_{\mathcal{T}}}{P_{\mathcal{R}}} , \tag{5}$$

which Planck put an upper bound $r \lesssim 0.11$ [3] while BICEP2 reported $r = 0.20^{+0.07}_{-0.05}$ [5].

So it seems that we are entering a new regime of observations using the primordial gravitational waves, although it is yet to be seen as of now how large is the fraction of the dust polarization in the BICEP2 result. If BICEP2 is mostly due to the primordial origin, the most favoured model is the simplest one with the potential $V(\phi) = m^2\phi^2/2$. Thus the energy scale of inflation is as high as 10^{16} GeV, with the field excursion being larger than m_{Pl}. Moreover, general relativity and quantum field theory seem to remain valid up to such a high energy scale, leading to the first clue of quantized gravity. But at the same time there seems no compelling hint beyond this simplest case. Thus the question we may naturally ask is what are the remaining windows for new physics beyond this simple picture.

2. Running of the primordial spectra

We first notice that there are a number of anomalies in the CMB. For example, there seems to exist hemispherical asymmetry between northern and southern hemisphere. On gravitational waves side, obviously there is a tension between Planck bound and BICEP2 detected value of r.

Among them, let us consider the low multipole regime of the TT power spectrum. Obviously the central values are at least marginally lower than the standard ΛCDM prediction [6]. Normalized at $k = 0.05$ Mpc^{-1}, the power on larger scales is suppressed than theoretical prediction. A possible solution is that the "running" of the scalar spectral index,

$$\alpha_{\mathcal{R}} \equiv \frac{dn_{\mathcal{R}}}{d\log k}, \tag{6}$$

is negative. This negative running is also supported to accommodate the large value of r: for Planck alone, the upper bound on r is relaxed to $r < 0.26$ with $\alpha_{\mathcal{R}} = -0.022 \pm 0.010$ [3]. To reconcile Planck and BICEP2, $\alpha_{\mathcal{R}} = -0.028 \pm 0.009$ [5] can do the job.

But in the standard, simplest model of inflation, $\alpha_{\mathcal{R}}$ is much suppressed. Expressed in terms of the slow-roll parameters,

$$\alpha_{\mathcal{R}} = -2\xi_V^2 - 24\epsilon_V^2 + 16\epsilon_V\eta_V, \tag{7}$$

where

$$\epsilon_V \equiv \frac{m_{\rm Pl}^2}{2}\left(\frac{V'}{V}\right)^2, \quad \eta_V \equiv m_{\rm Pl}^2\frac{V''}{V}, \quad \xi_V^2 \equiv m_{\rm Pl}^4\frac{V'V'''}{V^2}. \tag{8}$$

With $\epsilon_V = \mathcal{O}(0.01)$, $|\alpha_{\mathcal{R}}| = \mathcal{O}(\epsilon_V^2)$ and thus cannot accommodate the expected large value of $\alpha_{\mathcal{R}}$.

Because the scalar perturbation is dependent on the dynamics in the matter sector, we can obviously generalize to the case where inflation is driven by multiple number of fields to relax this difficulty. In the context of the so-called δN formalism [7], where the perturbation in the number of e-folds between the initial flat hypersurface during inflation and the final comoving one after inflation is equivalent to the final curvature perturbation, we can find [8]

$$\alpha_{\mathcal{R}} = 4\epsilon^2 - 2\epsilon\eta + \frac{N_aN_b}{N_dN^d}\left(8\epsilon w^{ab} + 4w^a{}_c w^{bc} - 2D_N w^{ab}\right) - (n_{\mathcal{R}}-1)^2, \tag{9}$$

where $N_a = \partial N/\partial\phi^a$, the slow-roll parameters are given in terms of the Hubble parameter as

$$\epsilon \equiv -\frac{\dot{H}}{H^2}, \quad \eta \equiv \frac{\dot{\epsilon}}{H\epsilon}, \tag{10}$$

and [9]

$$w_{ab} = u_{(a;b)} + \frac{R_{c(ab)d}}{3}\frac{\dot{\phi}_0^c}{H}\frac{\dot{\phi}_0^d}{H} \quad \text{with} \quad u_a = -\frac{V_{;a}}{3H^2}, \tag{11}$$

$$D_N w^{ab} = w^{ab}{}_{;c}\frac{\dot{\phi}_0^c}{H}. \tag{12}$$

In principle one can hope to accommodate large negative value of r in this case because the above general expression includes geometric information on the field space, like $R_{abcd;e}$ and $V_{;abc}$. But it seems to remain an open challenge to construct such a realistic model.

For the gravitational waves, there seem a number of anomalies, such as too high C_ℓ^{EE} and too low C_ℓ^{BB} at low multipoles. These may all be due to statistical uncertainty, but they may be real as well and we should wait for more accurate observations to be made very soon. But nevertheless these anomalies look quite interesting, and one of them is a positive bump at around $150 \lesssim \ell \lesssim 300$ beyond 2σ significance. This means the excess of the primordial gravitational waves more than $r = 0.20$ naively so that the spectral index of the tensor spectrum n_T is positive. But for standard case $n_T < 0$ always,

$$n_T = -2\epsilon. \tag{13}$$

Journal of Physics: Conference Series **627** (2015) 012012 doi:10.1088/1742-6596/627/1/012012

Since gravitational waves are after all gravity and thus only care the total energy density: they are blind to detailed dynamics during inflation, so considering multi-field inflation as we did for the scalar running does not help. Instead, we extend the notion of the slow-roll approximation where we implicitly assume hierarchies between a slow-roll parameter and its derivatives.

In this context of the "general slow-roll" formalism [10], the departure from perfect de Sitter phase is parametrized by a function $p(\log\tau) = -2\pi\tau a$, where $d\tau \equiv dt/a$ is the conformal time. Then the solution of the mode function can be written as a perturbative series of the Green's function solution [11]. Then, up to second order corrections n_T is written as [12]

$$n_T = 2\frac{p'}{p} + 2\alpha\left(\frac{p'}{p}\right)' + 2(4-\pi)\frac{p'p''}{p^2}\,, \tag{14}$$

where $p' \equiv dp/d\log(-\tau)$ and $\alpha \equiv 2 - \log 2 - \gamma \approx 0.577216$ with γ being the Euler-Mascheroni constant. Keeping up to this order, $n_T > 0$ as suggested by the bump in the BICEP2 result gives the condition $p'/p \gtrsim 1$ [13]. This can be translated into the slow-roll parameters as

$$\frac{d\log\epsilon}{d\log a} \lesssim -1\,, \tag{15}$$

so that ϵ is rapidly decaying as $1/a$ or even faster, so that η is negatively large. Of course, this period during which the above condition is satisfied cannot last too long, otherwise inflation does not terminate. Typical form of the potential in this case looks flat over a certain range, like ultra-slow-roll inflation [14] or punctuated inflation [15]. Of course, we may find a blue tilt of the tensor spectrum in different context, e.g. inflation driven by Galilion-type field [16] or string gas cosmology [17]. Also we may build other consistency checks, like the running of the tensor-to-scalar ratio [18], where we can eliminate ϵ contribution.

3. Effective single field inflation and features

In the previous section, we noticed the current observational data may suggest momentary deviations from otherwise vanilla predictions. Then one question remains ahead: the deviations, or more generally "features", mean there is a structure between smooth theory that spans super-Planckian field values. This is not surprising in the context of effective field theory: naively we expect smooth slow-roll inflation with large field excursion. But effective theory with a cutoff scale $\Lambda \lesssim m_{\rm Pl}$ allows in general higher dimensional operators $\mathcal{O}_n$ with the dimension $n > 4$ suppressed powers of Λ, such that

$$\mathcal{L}_{\rm eff}[\phi] = \mathcal{L}_0[\phi] + \sum_n c_n \frac{\mathcal{O}_n[\phi]}{\Lambda^{n-4}}\,, \tag{16}$$

where the coefficients c_n's are in general $\mathcal{O}(1)$. These higher dimensional terms bring sub-Planckian structure that generally prevents long enough slow-roll inflation without anything else, which gives deviations from otherwise smooth spectrum, viz. features. These higher dimensional operators can be obtained by two different approaches. One is to write down all the possible terms allowed by presumed symmetry principles, such as the Lorentz symmetry and gauge symmetry. In the popular approach of effective field theory of inflation, time translational symmetry is broken while spatial differmorphism is kept [19]. The other is, starting from a mother theory which contains degrees of freedom heavier than the scale of our interest, we integrate out those heavy modes. Here we concentrate on the latter, and see if there are any universal features of the integrated out heavy physics.

The recipe of obtaining effective single field theory is simple [20]. In the field space, the departure from the homogeneous and isotropic background solution $\phi_0^a(t)$ is represented by two fields $\pi = \pi(t, \boldsymbol{x})$ and $\mathcal{F} = \mathcal{F}(t, \boldsymbol{x})$ in such a way that

$$\phi^a(t, \boldsymbol{x}) = \phi_0^a(t + \pi) + N^a(t + \pi)\mathcal{F}\,, \tag{17}$$

where $N^a(t + \pi)$ stands for the normal vector to the background trajectory N^a evaluated at $t+\pi$. Notice that π represents deviations from $\phi_0^a(t)$ along the background trajectory, whereas $\mathcal{F}$ parametrizes deviations off the trajectory. For gravitational sector, we write the spatial metric γ_{ij} as

$$\gamma_{ij} = a^2(t + \pi)e^{2\mathcal{R}}\delta_{ij}\,. \tag{18}$$

Given that we will be eventually interested in the curvature perturbation $\mathcal{R}$ by integrating out $\mathcal{F}$, it is convenient to work in the comoving gauge where $\pi = 0$. Then, by performing the path integral over $\mathcal{F}$ we integrate out $\mathcal{F}$ and obtain the effective single field action of $\mathcal{R}$,

$$e^{S_{\text{eff}}[\mathcal{R}]} = \int [D\mathcal{F}]e^{S[\mathcal{R},\mathcal{F}]}\,, \tag{19}$$

which is at cubic order equivalent to plugging back the linear solution of $\mathcal{F}$ into the action [21],

$$\mathcal{F} = \left(-\Box + M_{\text{eff}}^2\right)^{-1} \frac{-2\dot{\theta}\dot{\phi}_0}{H}\dot{\mathcal{R}}\,, \tag{20}$$

where M_{eft} is the heavy mass scale and $\dot{\theta}$ is the angular velocity of the trajectory. The result is that the effects of heavy physics are described by the "speed of sound" given by

$$c_s^{-1} \equiv 1 + \frac{4\dot{\theta}^2}{M_{\text{eff}}^2}\,. \tag{21}$$

This is very intuitive: if the trajectory is not turning but straight where $\dot{\theta} = 0$, we can simply change the field space coordinates to have a single field description. If the mass scale M_{eff} is extremely heavy, $\mathcal{F}$ does not move at all and it can simply be ignored.

Being said "effective field theory", there must be a small parameter upon which we make the expansion of the theory. As can be read from the solution of $\mathcal{F}$, in the above prescription we have truncated at leading order in the expansion of $\Box/M_{\text{eff}}^2$,

$$\mathcal{F} = \frac{1}{M_{\text{eff}}^2}\left(1 + \frac{\Box}{M_{\text{eff}}^2} + \cdots\right)\frac{-2\dot{\theta}\dot{\phi}_0}{H}\dot{\mathcal{R}}\,. \tag{22}$$

Thus, the effective theory obtained by plugging back the leading solution remains valid for "adiabatic" trajectory for which $|\Box\mathcal{F}| \ll M_{\text{eff}}^2\mathcal{F}$, or [22]

$$\left|\frac{\ddot{\theta}}{\dot{\theta}}\right| \ll M_{\text{eff}}\,. \tag{23}$$

That is, the acceleration of $\mathcal{F}$ is small compared to the mass term so that the creation of heavy quanta is suppressed. Note that this means even for a extreme case where $c_s^{-2} \gg 1$ the effective theory is legitimate as long as the trajectory is adiabatic [23]. Of course, as the mass scale M approaches H the next-to-leading expansion term becomes important and cannot be ignored [24].

An interesting observational consequence is that if c_s sources features solely, the bispectrum is completely specified by the power spectrum and its first two derivatives, i.e. $n_{\mathcal{R}}$ and $\alpha_{\mathcal{R}}$.

Presuming that the reduction in the speed of sound is sufficiently small, by introducing a small parameter

$$u \equiv 1 - \frac{1}{c_s^2}, \tag{24}$$

we can find the change in the power spectrum generated by the speed of sound to first order in u as

$$\frac{\Delta\mathcal{P}_\mathcal{R}}{\mathcal{P}_\mathcal{R}}(k) = k\int_{-\infty}^{0} d\tau u(\tau)\sin(2k\tau), \tag{25}$$

where $\mathcal{P}_\mathcal{R} = 2\pi^2 P_\mathcal{R}/k^3$. By inverting this relation u can be written as a function of $\Delta\mathcal{P}_\mathcal{R}/\mathcal{P}_\mathcal{R}$. With the f_NL ansatz in mind, we define a dimensionless shape function as

$$f_\mathrm{NL}(k_1,k_2,k_3) \equiv \frac{10}{3}\frac{k_1k_2k_3}{k_1^3+k_2^3+k_3^3}\frac{(k_1k_2k_3)^2 B_\mathcal{R}}{(2\pi)^4\mathcal{P}_\mathcal{R}^2}. \tag{26}$$

Evaluating this shape function in certain interesting configurations, we find [25]

$$f_\mathrm{NL} = \begin{cases} \dfrac{5}{54}\left[-7\dfrac{\Delta\mathcal{P}_\mathcal{R}}{\mathcal{P}_\mathcal{R}} - 3\dfrac{d}{d\log k}\left(\dfrac{\Delta\mathcal{P}_\mathcal{R}}{\mathcal{P}_\mathcal{R}}\right) + \dfrac{d^2}{d\log k^2}\left(\dfrac{\Delta\mathcal{P}_\mathcal{R}}{\mathcal{P}_\mathcal{R}}\right)\right] & \left(\dfrac{k_2}{k_1} = \dfrac{k_3}{k_1} = 1 : \text{ equilateral}\right) \\ -\dfrac{5}{12}\dfrac{d}{d\log k}\left(\dfrac{\Delta\mathcal{P}_\mathcal{R}}{\mathcal{P}_\mathcal{R}}\right) & \left(\dfrac{k_2}{k_1} = 1, \dfrac{k_3}{k_1} \to 0 : \text{ squeezed}\right) \\ \dfrac{1}{8}\left[-\dfrac{\Delta\mathcal{P}_\mathcal{R}}{\mathcal{P}_\mathcal{R}} - \dfrac{5}{2}\dfrac{d}{d\log k}\left(\dfrac{\Delta\mathcal{P}_\mathcal{R}}{\mathcal{P}_\mathcal{R}}\right) + \frac{1}{2}\dfrac{d^2}{d\log k^2}\left(\dfrac{\Delta\mathcal{P}_\mathcal{R}}{\mathcal{P}_\mathcal{R}}\right)\right] & \left(\dfrac{k_2}{k_1} = 1, \dfrac{k_3}{k_1} = 2 : \text{ folded}\right) \end{cases}, \tag{27}$$

where $k = (k_1+k_2+k_3)/2$. This approach can be extended more generally when the features may be sourced by other origins [26], but still the bispectrum is specified by $\mathcal{P}_\mathcal{R}$, $n_\mathcal{R}$ and $\alpha_\mathcal{R}$.

4. Running of galaxy bias

Large scale structure (LSS) of the universe is, along with the CMB, yet another powerful cosmological probe, and its importance has ever been increasing with galaxy surveys such as SDSS. The LSS observations can provide the measurement of geometrical distances, growth of structures, and shape of primordial correlation functions. These lower redshift information combined with the CMB data can break down the degeneracies among cosmological parameters that yields better constraints than CMB alone. Furthermore, the full three-dimensional information with a huge redshift coverage available for the LSS observations naturally yields measurement of properties of dark energy, neutrino properties as well as physics of the early universe. A number of future observations such as MS-DESI, LSST and Euclid are planned to observe LSS with improved accuracy in near future.

What we measure in large scale survey are galaxies. But as is well known, significant fraction of matter in the universe is dark matter while galaxies consist of baryons. Thus, the distribution of galaxies is not precisely the same as that of dark matter. Thus, the galaxy density field, $\delta_g \equiv (n_g - \bar{n}_g)/\bar{n}_g$, is different from the density perturbation of dark matter, $\delta \equiv (\rho_m - \bar{\rho}_m)/\bar{\rho}_m$. This discrepancy is parametrized by the so-called bias factor b,

$$\delta_h = b\delta. \tag{28}$$

Thus the question is how to improve our theoretical prediction of b, i.e. refining our arguments of identifying the locations of the galaxy formation, i.e. density peaks in the matter density distribution, which follows nearly Gaussian probability distribution. An obvious approach is that we identify peaks as the regions where δ exceeds a certain threshold density $\delta_c = \nu\sigma_0$ with $\nu \gg 1$, where σ_0 is the first spectral moment,

$$\sigma_n^2 \equiv \int \frac{d^3k}{(2\pi)^3} P(k) k^{2n} \,. \tag{29}$$

The bias factor can be computed by functional integration and is found to be a constant. Including primordial non-Gaussianity we find a non-trivial scale dependence proportional to $k^{-2} f_{\rm NL}$ [27] and $k^{-4}\left[\tau_{\rm NL} - (6f_{\rm NL}/5)^2\right]$ [28].

A more refined approach is that we identify peaks as local maxima of δ [29]. This is reasonable since after non-linear evolution such local peaks will be enhanced and will eventually lead to the formation of galaxies. To define a peak, we need to specify the density field itself and its first and second derivatives,

$$\nu \equiv \frac{\delta}{\sigma_0} \,, \quad \eta_i \equiv \frac{\partial_i\delta}{\sigma_1} \,, \quad \zeta_{ij} \equiv \frac{\partial_i\partial_j\delta}{\sigma_2} \,, \tag{30}$$

where the minimum eigenvalue of the 3×3 matrix $-\zeta_{ij}$ is greater than 0. Then we can proceed further to include corrections from the Gaussian cross-correlation between two points $\boldsymbol{x}_1$ and $\boldsymbol{x}_2$, and non-Gaussian corrections from 3-point correlation function. The former is obtained by simply expanding the 20×20 covariance matrix [30]

$$\mathcal{M} = \begin{array}{c} \\ \boldsymbol{y}(\boldsymbol{x}_1) \\ \boldsymbol{y}(\boldsymbol{x}_2) \end{array} \!\!\begin{array}{c} \begin{array}{cc} \boldsymbol{y}(\boldsymbol{x}_1) & \boldsymbol{y}(\boldsymbol{x}_2) \end{array} \\ \begin{pmatrix} M & B \\ B^T & M \end{pmatrix} \end{array} , \tag{31}$$

where M and B are 10×10 self- and cross-correlation matrices respectively and at each point we construct a 10-dimensional vector $\boldsymbol{y}^T = (\eta_i \; \nu \; \zeta_{ij})$. Then the Gaussian exponent can be expanded as a series of B as

$$\exp\left(-\frac{1}{2}\boldsymbol{y}^T \mathcal{M}^{-1} \boldsymbol{y}\right) = \left[1 + \boldsymbol{y}^T(\boldsymbol{x}_1) M^{-1} B M^{-1} \boldsymbol{y}(\boldsymbol{x}_2) + \cdots\right] \exp\left(-\sum_i \frac{1}{2}\boldsymbol{y}^T(\boldsymbol{x}_i) M^{-1} \boldsymbol{y}(\boldsymbol{x}_i)\right) . \tag{32}$$

The non-Gaussian corrections consist of 8 non-zero contributions of 3-point correlation functions,

$$\begin{aligned} & \langle\nu(1)\nu(1)\nu(2)\rangle \,, \quad \langle\nu(1)\nu(1)\zeta_{ij}(2)\rangle \,, \quad \langle\nu(1)\zeta_{ij}(1)\nu(2)\rangle \,, \quad \langle\nu(1)\zeta_{ij}(1)\zeta_{kl}(2)\rangle \,, \\ & \langle\eta_i(1)\eta_j(1)\nu(2)\rangle \,, \quad \langle\eta_i(1)\eta_j(1)\zeta_{kl}(2)\rangle \,, \quad \langle\zeta_{ij}(1)\zeta_{kl}(1)\nu(2)\rangle \,, \quad \langle\zeta_{ij}(1)\zeta_{kl}(1)\zeta_{mn}(2)\rangle \,. \end{aligned} \tag{33}$$

The combined non-Gaussian corrections lead to non-monotonic running of the bias factor [31].

5. Conclusions

To summarize, most recent observations are consistent with simplest models of inflation, such as large field models with power-law type potential. But nevertheless there are various windows to probe beyond this simplest picture: we have listed the scale of ultraviolet physics through the correlated features in the correlation functions, the dynamics of inflation via various runnings of the primordial spectra, and non-linear processes by large scale galaxy clustering. More importantly, we have a number of ongoing and planned observations ahead. Very soon we will have more accurate data on the CMB polarization from Planck, Keck Array and BICEP3. Furthermore, next generation experiments of the galaxy survey such as DESI and Euclid will be operational within next decade. Thus we are indeed enjoying the most exciting moment in physical cosmology and we must be expecting the unexpected.

Acknowledgments

I thank the organizers of the 2nd International Workshop on Theoretical and Computational Physics in Buon Ma Thuot, Vietnam during 28 - 31 July, 2014. I acknowledge the Max-Planck-Gesellschaft, the Korea Ministry of Education, Science and Technology, Gyeongsangbuk-Do and Pohang City for the support of the Independent Junior Research Group at the Asia Pacific Center for Theoretical Physics. I am also supported by a Starting Grant through the Basic Science Research Program of the National Research Foundation of Korea (2013R1A1A1006701).

References

[1] A. H. Guth, Phys. Rev. D **23**, 347 (1981) ; A. D. Linde, Phys. Lett. B **108**, 389 (1982) ; A. Albrecht and P. J. Steinhardt, Phys. Rev. Lett. **48**, 1220 (1982).
[2] See e.g. D. H. Lyth and A. R. Liddle, "The primordial density perturbation: Cosmology, inflation and the origin of structure," Cambridge, UK: Cambridge Univ. Pr. (2009) 497 p
[3] P. A. R. Ade *et al.* [Planck Collaboration], arXiv:1303.5082 [astro-ph.CO].
[4] P. A. R. Ade *et al.* [Planck Collaboration], arXiv:1303.5084 [astro-ph.CO].
[5] P. A. R. Ade *et al.* [BICEP2 Collaboration], Phys. Rev. Lett. **112**, 241101 (2014) [arXiv:1403.3985 [astro-ph.CO]].
[6] P. A. R. Ade *et al.* [Planck Collaboration], Astron. Astrophys. (2014) [arXiv:1303.5076 [astro-ph.CO]].
[7] M. Sasaki and E. D. Stewart, Prog. Theor. Phys. **95**, 71 (1996) [astro-ph/9507001] ; J. -O. Gong and E. D. Stewart, Phys. Lett. B **538**, 213 (2002) [astro-ph/0202098].
[8] J. O. Gong, arXiv:1409.8151 [astro-ph.CO].
[9] J. Elliston, D. Seery and R. Tavakol, JCAP **1211**, 060 (2012) [arXiv:1208.6011 [astro-ph.CO]] ; C. T. Byrnes and J. O. Gong, Phys. Lett. B **718**, 718 (2013) [arXiv:1210.1851 [astro-ph.CO]].
[10] E. D. Stewart, Phys. Rev. D **65**, 103508 (2002) [astro-ph/0110322] ; J. Choe, J. -O. Gong and E. D. Stewart, JCAP **0407**, 012 (2004) [hep-ph/0405155].
[11] J. O. Gong and E. D. Stewart, Phys. Lett. B **510**, 1 (2001) [astro-ph/0101225].
[12] J. -O. Gong, Class. Quant. Grav. **21**, 5555 (2004) [gr-qc/0408039].
[13] J. O. Gong, JCAP **1407**, 022 (2014) [arXiv:1403.5163 [astro-ph.CO]].
[14] N. C. Tsamis and R. P. Woodard, Phys. Rev. D **69**, 084005 (2004) [astro-ph/0307463] ; W. H. Kinney, Phys. Rev. D **72**, 023515 (2005) [gr-qc/0503017].
[15] R. K. Jain, P. Chingangbam, J. -O. Gong, L. Sriramkumar and T. Souradeep, JCAP **0901**, 009 (2009) [arXiv:0809.3915 [astro-ph]].
[16] T. Kobayashi, M. Yamaguchi and J. Yokoyama, Phys. Rev. Lett. **105**, 231302 (2010) [arXiv:1008.0603 [hep-th]].
[17] R. H. Brandenberger, A. Nayeri, S. P. Patil and C. Vafa, Phys. Rev. Lett. **98**, 231302 (2007) [hep-th/0604126] ; R. H. Brandenberger, A. Nayeri and S. P. Patil, arXiv:1403.4927 [astro-ph.CO].
[18] J. -O. Gong, Phys. Rev. D **79**, 063520 (2009) [arXiv:0710.3835 [astro-ph]].
[19] C. Cheung, P. Creminelli, A. L. Fitzpatrick, J. Kaplan and L. Senatore, JHEP **0803**, 014 (2008) [arXiv:0709.0293 [hep-th]].
[20] A. Achucarro, J. -O. Gong, S. Hardeman, G. A. Palma and S. P. Patil, JHEP **1205**, 066 (2012) [arXiv:1201.6342 [hep-th]].
[21] A. Achucarro, J. -O. Gong, S. Hardeman, G. A. Palma and S. P. Patil, JCAP **1101**, 030 (2011) [arXiv:1010.3693 [hep-ph]].
[22] S. Cespedes, V. Atal and G. A. Palma, JCAP **1205**, 008 (2012) [arXiv:1201.4848 [hep-th]].
[23] A. Achucarro, V. Atal, S. Cespedes, J. O. Gong, G. A. Palma and S. P. Patil, Phys. Rev. D **86**, 121301 (2012) [arXiv:1205.0710 [hep-th]].
[24] J. O. Gong, M. S. Seo and S. Sypsas, arXiv:1407.8268 [hep-th].
[25] A. Achucarro, J. -O. Gong, G. A. Palma and S. P. Patil, Phys. Rev. D **87**, 121301 (2013) [arXiv:1211.5619 [astro-ph.CO]].
[26] J. -O. Gong, K. Schalm and G. Shiu, Phys. Rev. D **89**, 063540 (2014) [arXiv:1401.4402 [astro-ph.CO]].
[27] N. Dalal, O. Dore, D. Huterer and A. Shirokov, Phys. Rev. D **77**, 123514 (2008) [arXiv:0710.4560 [astro-ph]].
[28] J. O. Gong and S. Yokoyama, Mon. Not. Roy. Astron. Soc. **417**, 79 (2011) [arXiv:1106.4404 [astro-ph.CO]].
[29] J. M. Bardeen, J. R. Bond, N. Kaiser and A. S. Szalay, Astrophys. J. **304**, 15 (1986).
[30] V. Desjacques, Phys. Rev. D **78**, 103503 (2008) [arXiv:0806.0007 [astro-ph]].
[31] V. Desjacques, J. O. Gong and A. Riotto, JCAP **1309**, 006 (2013) [arXiv:1301.7437 [astro-ph.CO], arXiv:1301.7437].

Double ionization of molecule H_2 in intense ultrashort laser fields

Thu-Thuy Le, Ngoc-Ty Nguyen

Department of Physics, Ho Chi Minh University of Pedagogy, 280 An Duong Vuong Street, District 5, Ho Chi Minh City, Vietnam

E-mail: tynn@hcmup.edu.vn

Abstract. By solving numerically the time-dependent Schrödinger equation (TDSE), we have calculated the double ionization probability when a vibrating hydrogen molecule interacts with intense ultrashort laser pulses. The results show that in the case of vibrating nuclei the double ionization probability is higher than that of the fixed nuclei. Additionally, the double ionization probability is larger if the molecule is vibrating in a higher level. This is due to the decreasing of ionization potential when the inter-nuclei separation increases.

1. Introduction

Since intense ultrashort laser fields were produced, interaction between them and atoms, molecules has attracted much interests. When an atom or molecule is exposed to an intense laser field, several non-linear phenomena may occur, e.g. ionization, high-order harmonic generation. Single ionization has been studied for one-electron atoms [1, 2] but with many-electron atoms or molecules effects due to the other electrons or molecular structure need to be involved.

With many-electron atoms, ionization process becomes more complex. In this paper, we focus on double ionization of hydrogen molecule. There are two mechanisms: sequential double ionization and non-sequential double ionization. Sequential double ionization is a process of formation of doubly-charged ions consisting of two single-electron ionization events: the first electron is removed from a neutral atom or molecule followed by detachment of the second electron from the ion. Non-sequential double ionization is a process whose mechanism differs from the sequential one. For example, both the electrons leave the system simultaneously or the second electron's liberation is assisted by the first electron [3, 4].

Investigation of double ionization has been carried out widely with many different methods. In [5], Chen *et al.* calculated momentum distribution of recoil ions He^{2+} from laser-induced non-sequential double ionization of helium by a semi-classical rescattering model. Moreover, Kopold *et al.* also developed a method for calculation of the S matrix applied to the distribution for the total electronic momentum. They claimed that the result well agreed with measurement for neon but there was no comparable agreement for helium and argon [6]. In [7] Winger transformation is used to study the electronic center-of-mass motion in phase space for double ionization in a strong laser field. Continuously, in 2006 Parker *et al.* measured electron energy distributions of singly and doubly ionized helium in an intense 390 nm laser field at two intensities 0,8 PW/cm^2 and 1,1 PW/cm^2. Authors also confirmed that their numerical solution from solving the time-dependent Schrödinger equation (TDSE) was consistent with measurement [8]. In addition, in [9] Denys

I. Bondar *et al.* developed an analytical method for correlated two-electron in strong infrared laser fields. Authors pointed out that if two electrons are ejected nearly simultaneously, they are driven by the laser field in the same direction. If two electron are ejected with a substantial delay, the quarter-cycle or more, they end up going in the opposite directions.

Double ionization of H_2, the simplest molecule has been studied, is researched by many groups [10–13]. In [10] Volkova *et al.* calculated the ratio of single and double ionization of the hydrogen molecule to find out the mechanism of double ionization including tunneling and multiphoton. Furthermore, Sagout *et al.* set up experiment to measure and calculated the proton spectrum as a function of the pulse laser, the carrier-envelope and the intensity [11]. The dependence of double ionization yield of H_2 on carrier-envelope phase was also reported in [12]. Li Yan *et al*, in addition, studied probabilities of double ionization of H_2 for parallel and perpendicular alignment. They concluded that the probability is higher if the molecule is aligned parallel [13].

However, in above mentioned papers, calculations are only carried out with frozen nuclei. Nuclear vibration, in our point of view, should be taken into account. Here, we solve numerically the TDSE to investigate the influence of vibrational levels on double ionization probabilities with different laser parameters. The rest of the paper is arranged as follows. In Sec. 2, we introduce a numerical method to calculate the double ionization probability for H_2. In Sec. 3, we show our results of calculation for the probabilities with fixed and vibrating nuclei. The effects of different vibration levels are also analysed in this section. Sec. 4 is the conclusions.

2. Model and basic formalism

In this work, one-dimensional model H_2 is used and we solve numerically the TDSE to obtain the double ionization probability. Atomic units are used throughout the paper unless stated.

The TDSE for a H_2 exposed to an intense laser field can be written as

$$i\frac{\partial}{\partial t}\Psi\left(x_1,x_2,R,t\right)=\left(-\frac{\partial^2}{2\partial x_1^2}-\frac{\partial^2}{2\partial x_2^2}-\frac{\partial^2}{2\mu\partial R^2}+V_C\left(x_1,x_2R\right)+V_L\left(x_1,x_2,t\right)\right)\Psi\left(x_1,x_2,R,t\right) \tag{1}$$

where x_1, x_2 are electrons coordinate with respect to nuclear center of mass; R is the internuclear separation; μ is the reduced mass of two nuclei; $V_C\left(x_1,x_2,R\right)$ is a soft-core Coulomb potential and $V_L\left(x_1,x_2,t\right)$ is the laser-induced potential. In this model, the soft-core Coulomb potential in equation (1) can be written as

$$\begin{aligned}V_C\left(x_1,x_2,R\right)=&-\frac{1}{\sqrt{\left(x_1-\frac{R}{2}\right)^2+a}}-\frac{1}{\sqrt{\left(x_1+\frac{R}{2}\right)^2+a}}-\frac{1}{\sqrt{\left(x_2-\frac{R}{2}\right)^2+a}}\\&-\frac{1}{\sqrt{\left(x_2+\frac{R}{2}\right)^2+a}}+\frac{1}{\sqrt{\left(x_1-x_2\right)^2+a}}+\frac{1}{R},\end{aligned} \tag{2}$$

Constant a called soft-core parameter is added to avoid the singularity of the Coulomb potential and to mimic the real potential energy curve (PEC) of H_2. In this paper a is considered to depend parametrically in the internuclear separation. The laser-induced potential is given by

$$V_L\left(x_1,x_2,t\right)=\left(x_1+x_2\right)E_0f(t)\cos\left(\omega_0t\right) \tag{3}$$

where $E_0, f(t), \omega_0$ are the electric field amplitude, envelop function and frequency of lase pulse, respectively. The solution for equation (1) can be written as follows

$$\Psi\left(x_1,x_2,R,t\right)=\exp\left[-i\int_0^t\hat{H}dt\right]\Psi\left(x_1,x_2,R,0\right) \tag{4}$$

where $\Psi(x_1, x_2, R, 0)$ is the initial wave function which is given by the imaginary relaxation method. We also employ the Born-Oppenheimer approximation to obtain the wave function of the molecule. The split operator is used to obtain the time-dependent wave function in (4) [14]. With the time-dependent wave function, we calculate the probability of single and double ionization by

$$P(V,t) = \int_V |\Psi(x_1, x_2, R, t)|^2 dx_1 dx_2 dR. \quad (5)$$

where V is the integral region. Double ionization occurs in the asymptotic region $V_2 \equiv \{|x_1| \geq x_a, |x_2| \geq x_a\}$. Single ionization region is defined as $V_1 \equiv \{|x_1| < x_a, |x_2| > x_a \cap |x_2| < x_a, |x_1| > x_a\}$, and the neutral molecule is found in $V_0 \equiv \{|x_1|, |x_2| \leq x_a\}$ with $x_a = 20a.u.$

3. Results

In this section, we show calculation results of the single and double ionization of H_2 in intense laser fields. First, nuclei are fixed to investigate the dependence of probabilities on laser parameters such as intensity and pulse duration. In Fig.1 the single and double ionization probability of H_2 in a laser field with wave length of 1200nm, intensity of 3×10^{14}W/cm^2 and 10-cycle pulse is demonstrated. One can see that from 1st to 3rd cycle, single ionization probability is negligible, from 3rd to 7th the probability increases rapidly and is followed by a plateau region after 7th cycle. In this case, the maximum probability is 36,6%. The double ionization probability also has the similar performance with smaller values. In Fig.1 the double ionization only can be considered to occur from 5th cycle and the maximum value 7,6% is, five times smaller than the single ionization probability.

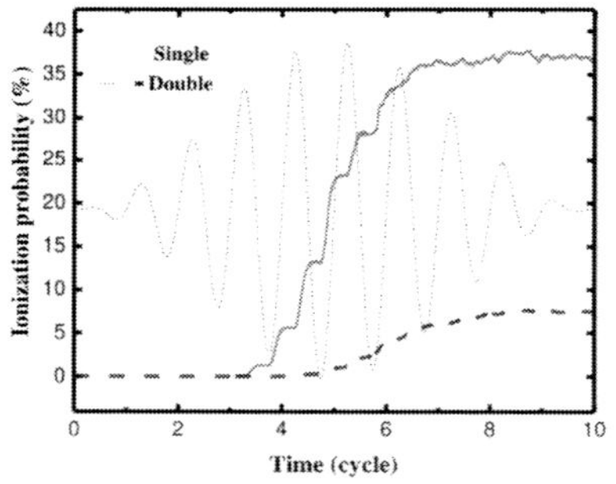

Figure 1. The single ionization (solid line) and double ionization probability (dashed line) of H_2 in a laser field with peak intensity of 3×10^{14}W/cm^2, wave length of 1200nm and duration of 40fs (10 cycles).

Thus, single ionization may be considered to occur from 3rd cycle while double ionization happens at 5th cycle. Both processes reach the maximum value at 7th cycle which is followed by a plateau region.

Those results can be explained by using the three-step model. After tunneling, the 1st electron will be accelerated by electric field. When the electric field reverses, the electron will be pulled back and collide with the 2nd electron that leads to double ionization. That is the reason why the double ionization happens nearly two cycles later than single ionization.

In Fig.1, there are some hollows in the curves of ionization probabilities. That is due to being driven back of electrons into neutral molecule region making the ionization probabilities smaller. When releasing electrons balances with returning, the probabilities reach the saturated value.

Next, we calculate the double ionization probability of H_2 with vibrating nuclei. The results show that the probability of a H_2 with vibrating nuclei is higher than that with fixed ones. In Fig.2, we plot the probabilities of H_2 exposed to a laser field with peak intensity of $2\times10^{14}\text{W/cm}^2$, wave length of 1200nm and duration of 16fs.

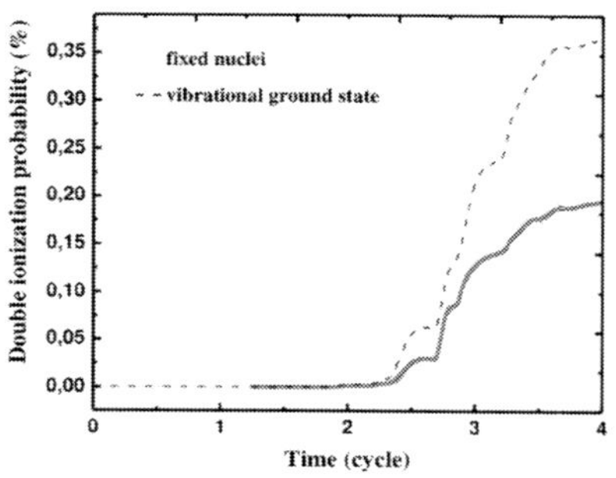

Figure 2. Double ionization probability of fixed nuclei H_2 (solid line) and H_2 in vibrational ground state (dashed line) in a laser field with peak intensity of $2 \times 10^{14}\text{W/cm}^2$, wave length of 1200nm and duration of 16fs (4 cycles).

With the vibrating molecule, the probability also has a similar behaviour to that of the fixed one but about 1,85 times higher amplitude, i.e., 0.37% and 0,2% respectively. These results may be explained by using terms of ionization potential energy. The energy of a vibrating molecule is always higher than that of a frozen one so the ionization potential is lower. We continue to calculate the probabilities with different laser's parameters and we also obtain the similar results.

Finally, we investigate the effects of vibration levels on the double ionization probabilities. The vibration levels are limited from $\nu = 0$ to $\nu = 5$. In Fig.3, we show the double ionization probabilities of a vibrating H_2 in different levels exposed to a laser pule with peak intensity of $2 \times 10^{14}\text{W/cm}^2$, wave length of 1200 nm and duration of 16fs.

In Fig. 3, one can see that the probability from a higher vibration level is larger than that from a lower one. The reason is because the ionization potential of the molecule in a lower vibration level is larger than that quantity of a higher one. That leads to a smaller probability to tunnel of electrons. We continue to calculate with different laser's parameters, the similar phenomena are also obtained. Thus, we conclude that double ionizing atoms in higher vibration levels occurs with larger probabilities.

4. Conclusions

In this paper, we solve numerically the TDSE for 1D H_2 in intense laser fields to calculate double ionization probabilities. The results point out that the probabilities of the vibrating molecule is always higher than that of a fixed one. In addition, if the molecule is in a higher vibration level, the probability is larger. These results can be explained by using terms of the ionization potential. In next works, extending the method for 2D or 3D molecules need to be investigated.

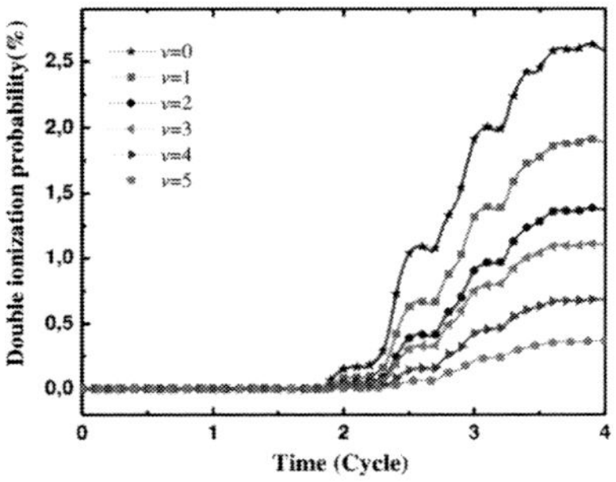

Figure 3. Double ionization probabilities of H_2 with different vibration levels. The laser field with peak intensity of $2 \times 10^{14}\mathrm{W/cm^2}$, wave length of 1200 nm and duration of 16fs is used.

Acknowledgments
This research is funded by Vietnam National Foundation for Science and Technology Development (NAFOSTED) under grant number 103.01-2013.37 and HCM University of Pedagogy, grant no. 14.19.63.

References
[1] Popov A M, Tikhonova O V, and Volkova E A 1999 *Lase. Phys.* **9** 1053
[2] Thriumalai A and Heyl J S 2009 *Phys. Rev. A* **79** 012514
[3] Baier S and Becker A 2008 *Phys. Rev. A* **78** 013409
[4] Baier S, Ruiz C, Plaja L, and Becker A 2006 *Phys. Rev. A* **74** 033405
[5] Chen J, Liu J, Fu B L and Zheng M W 2000 *Phys. Rev. A* **63** 011404(R)
[6] Kopold R, Becker W, Rottke H, and Sandner W. 2000 *Phys. Rev. Lett.* **85** 3781
[7] Lein M, Engel V, Gross E K U 2001 *Opt. Exp.* **8** 441
[8] Parker J S, Doherty B J S, Taylor K T, Schultz K D, Blaga C I, and DiMauro L F 2006 *Phys. Rev. Lett.* **96** 133001
[9] Bondar Denys I, Liu Wing-Ki, Ivanov Misha Yu 2009 *Phys. Rev. A* **79** 023417
[10] Volkova E A, Popov A M, and Tikhonova O V 2002 *Lase. Phys.* **13** 435
[11] Saugout S, Charron E, Cornaggia C 2008 *Phys. Rev. A* **77** 023404
[12] Tong X M and Lin C D 2007 *J. Phys. B* **40** 641
[13] Yan Li, Shi-PingYang, Xin-Yan Jia, and Jing Chen 2010 *Chin. Phys. B* **19** 043303
[14] Feit M D, Jr Fleck J A, and Steiger A 1982 *J. Comp. Phys.* **47** 412

Effects of nuclear vibration on the ionization process of H_2^+ in ultrashort intense laser field

Ngoc-Loan Phan[1], **Tran-Chau Truong**[2], **Ngoc-Ty Nguyen**[1]

[1] Department of Physics, Ho Chi Minh University of Pedagogy, 280 An Duong Vuong Street, District 5, Ho Chi Minh City, Vietnam

[2] Faculty of Foundation Sciences, Ho Chi Minh City University of Technology and Education, 1 Vo Van Ngan Street, Thu Duc District, Ho Chi Minh city, Vietnam

E-mail: tynn@hcmup.edu.vn

Abstract. By numerically solving the time-dependent Schrödinger equation, we calculate the ionization probability of a vibrating H_2^+ exposed to ultrashort intense laser fields. The results show that the ionization probability increases by time and gets a saturation value. We also find that with some first vibration levels, the ionization probability from a higher vibration level is larger than that from a lower one. However, with higher vibration levels, at a certain level the ionization probability will take maximum and decrease with next levels.

1. Introduction

Since the first optical laser source was invented by Theodore Maiman in 1960 [1], the laser technology has not only grown sharply but also has promoted the scientific and technical revolution. Beside efforts to shorten the pulse duration and to increase the intensity of the laser fields, scientists are especially concerned about the laser-matter interaction. The interaction of ultrashort intense laser pulses with atoms or molecules leads to many nonlinear optical phenomena such as high harmonic generation (HHG) [2–4], above threshold dissociation (ATD), and above threshold ionization (ATI) [5,6].

The ionization process plays an important role in atom, molecule-laser interaction because it is the background to explain above mentioned phenomena. Many works have been conducted to find out ionization mechanisms, specially Keldyshs article published in 1965 [7]. According to him, there are three mechanisms of ionization of atoms, molecules including tunneling, multiphoton, and above threshold ionization. In fact, which mechanism is dominant depends on frequency, peak intensity of laser fields, and the ionization potential of atoms as well. Unlike atoms, the problems related to interaction between laser and molecules are considerably complex because of a large number of variables. While the problem of intense laser-atom interaction is solved almost completely by using the single active electron approximation (SAE) [2,4], the scientists continue to study the molecular systems in electromagnetic fields [8–11]. However, due to many degrees of freedom, finding wave functions of molecules and other quantities such as the ionization probability, ionization rate are much more difficult than those for atoms. By using the semi-classical three-step model, the amount of calculation significantly reduces [2]. On the other hand, the numerically solving the time dependent Schrdöinger equation (TDSE) approach can give exact results but requires large computer resource, so now this method is only applied to some simple molecules such as H_2, H_2^+, H_3^{2+}.

To simplify the laser-molecule interaction, the nuclei are assumed to be fixed during interaction process and their vibrations are not taken into account when calculating the ionization rate and HHG [3]. However, recently scientists have pointed out many differences between two cases of frozen and moving nuclei in both the HHG spectra and the ionization rate [8–11]. In [8], authors claim that when the initial nuclear vibration is assumed in the ground state, the ionization probabilities and the internuclear distance of molecular ion H_2^+ increase in compared with that in the case of fixed nuclei. Ionization rate of H_2^+ for different initial vibrational states $\nu = 0 \div 3$ is also studied in [9]. The authors confirm that the faster nuclei vibrate, the higher ionization rate is. Moreover, single and double ionization probabilities of the H^2 molecule are calculated to explain the changes of the HHG spectra due to influence of nuclear vibration [10]. In [8–10], ionization rate is only calculated with few low vibration states of nuclei, the effects of high nuclear vibration levels to the ionization process have not considered yet. In this article, we calculate the ionization probabilities of H_2^+ when two nuclei are prepared in various initial vibrational states $\nu = 0 \div 21$ before the molecule interacts with intense laser fields. In addition, information of nuclear dynamic such as internuclear distance, relative velocity of the two nuclei is also studied to explain ionization related results.

The rest of the paper is arranged as follows. In Sec. 2, we introduce a numerical method to calculate the ionization probability for H_2^+. In Sec. 3, we show our results of calculation for the probabilities with fixed and vibrating nuclei. The effects of different vibration levels are also analysed in this section. Sec. 4 is the conclusions.

2. Calculation Method

When a molecular ion H_2^+ interacts with a linearly polarized laser field, after being ionized, electron moves along the laser polarization direction. On the other hand, as an electric dipole in an electric field, the molecule is quickly aligned along the polarization vector of the intense laser field. Thus we restrict our study to the one-dimensional model (1D) to reduce the amount of calculations. In this section, we introduce a method to solve the TDSE of a vibrating H_2^+ exposed to intense laser fields. After obtaining the time-dependent wave function, we calculate the ionization probabilities and nuclear separation. We use the atomic units ($\hbar = e = m_e = 1$) throughout the paper unless stated. The TDSE of the H_2^+ interacting with an intense pulsed laser has the following form

$$i\frac{\partial}{\partial t}\Psi(z,R,t) = \left(-\frac{\partial^2}{2\partial z^2} - \frac{\partial^2}{2\mu\partial R^2} + V_C(z,R) + V(z,t)\right)\Psi(z,R,t), \tag{1}$$

where z is the electrons coordinate respect to the center of mass of two nuclei, R is the internuclear distance, μ is the reduced mass of the two nuclei. $V_C(z,R)$ is the soft Coulomb potential written as

$$V_C(z,R) = \frac{1}{R} - \frac{1}{\sqrt{\left(z-\frac{R}{2}\right)^2+\varepsilon}} - \frac{1}{\sqrt{\left(z+\frac{R}{2}\right)^2+\varepsilon}}. \tag{2}$$

The constant ε, called the softening parameter, is added to the Coulomb potential to avoid the singularities. We choose $\varepsilon = 1$ so that the ground state energy in one-dimensional model can mimic the real potential energy curve [6, 11]. With this model, the equilibrium internuclear distance of molecular ion H_2^+ is obtained as $R = 2.609\,a.u.$ The laser-molecule interaction is formulated in length gauge as

$$V(z,t) = zE_0 f(t)\sin(\omega_0 t), \tag{3}$$

where E_0, ω_0 are the electric amplitude and the angular frequency of laser pulse, respectively. $f(t)$, called the envelope of the laser pulse, has a sine-squared shape. The TDSE (1) can be numerically solved by using the split operator method [12]. By that way, the wave function $\Psi(z, R, t)$ is calculated by applying the time evolution operator to the initial wave function $\Psi(z, R, t = 0)$ as follows

$$\Psi(z, R, t) = \exp\left[-i \int_0^t \hat{H} dt\right] \Psi(z, R, t = 0). \tag{4}$$

In this study, the approximation BO is utilized in order to calculate the initial wave function. With each nuclear vibration level, characterized by the quantum number ν, the wave function is set as

$$\Psi(z, R, t = 0) = \varphi\left(z, \overline{R}\right) \chi_v(R), \tag{5}$$

where $\varphi\left(z, \overline{R}\right)$, $\chi_v(R)$ are the electronic and vibration nuclear wave functions, respectively. The electronic wave function $\varphi\left(z, \overline{R}\right)$ which can be solved by the imaginary relaxation method [13] is first calculated from the Schrödinger equation of electronic system at different values of internuclear R as follows

$$\left(-\frac{\partial^2}{2\partial z^2} + V_C\left(z, \overline{R}\right)\right) \varphi\left(z, \overline{R}\right) = E_e\left(\overline{R}\right) \varphi\left(z, \overline{R}\right). \tag{6}$$

The nuclear wave function $\chi_v(R)$ is also determined by the same method in order to solve the Schrödinger equation of a particle with the reduced mass, bounded in the electronic potential as

$$\left(-\frac{\partial^2}{2\mu \partial R^2} + E_e(R)\right) \chi_v(R) = E \chi_v(R). \tag{7}$$

After obtaining the wave function of the molecule, we continue to calculate the ionization probability. We define the ionization region S_i as

$$S_i = \{z | Abs(z) > z_i + R/2\}, \tag{8}$$

where $z_i = 20\, a.u.$ The time-dependent ionization probabilities can be calculated as

$$P_i(t) = \int\int_{S_i} |\Psi(z, R, t)|^2 dz dR. \tag{9}$$

Moreover, we also calculate some other quantities, relating to nuclear and electronic dynamics. Concerning the relative dynamics of the nuclei, the internuclear separation is expressed as

$$R(t) = \int\int R|\Psi(z, R, t)|^2 dz dR. \tag{10}$$

Also, the relative velocity between the two nuclei is calculated from the expectation value of the nuclear distance by the following formula $\vartheta(t) = dR/dt$.

In this paper, the numerical grid points are chosen over $0.5\, a.u. \leq R \leq 20.5\, a.u.$ and $-800\, a.u. \leq z \leq 800\, a.u.$ at interval of $\Delta z = 0.39\, a.u.$, $\Delta R = 0.04\, a.u.$ and the time step $\Delta t = 0.18\, a.u.$

Journal of Physics: Conference Series **627** (2015) 012014 doi:10.1088/1742-6596/627/1/012014

3. Results

In this section, we show our results about effect of molecular vibration on the ionization probabilities. We calculate the time-dependent ionization probabilities of the H_2^+ interacting with ultrashort laser pulses in both cases of fixed and moving nuclei at different vibration states. The laser with wave length of 800 nm, pulse duration of 26 fs (10 cycles), peak intensity of $3 \times 10^{14}\, W/cm^2, 8 \times 10^{14}\, W/cm^2$ are used. The results with peak intensity of $3 \times 10^{14}\, W/cm^2$ are shown in Fig. 1a, which indicate that the onset of ionization occurs after about three optical laser cycles. The ionization probabilities significantly increase in the next cycles, then almost unchanged in the last few ones. In Fig. 1b, the calculation results with peak intensity of $8 \times 10^{14}\, W/cm^2$ are presented. In Fig. 1, we show that when the laser is turned off, the ionization probabilities rapidly enhance at some first vibration levels, then slightly increase at higher vibration levels. The ionization probability of the H_2^+ in the fields of laser pulses with peak intensity of $3 \times 10^{14}\, W/cm^2$, wavelength of 800 nm and pulse duration of 26 fs is quite small ($\sim 4.2\%$) in case of moving nuclei at the ground vibrational state ($v = 0$). However, ionization probabilities strongly increase and equal 22% when nuclei vibrate at $v = 1$; at $v = 2$, they increase to 42% and reach to 51% when $v = 3$ (Fig. 1a). With higher laser intensity, the onset time of ionization sooner occurs and the differences of ionization probabilities (when the laser switch off) between vibration levels are smaller. When the molecule exposed to the laser with intensity of $8 \times 10^{14}\, W/cm^2$, the ionization probabilities reach to 94% in case of moving nuclei at ground level ($v = 0$), and nearly unchanged when nuclei oscillate at higher vibration states (Figure 1b). The ionization probabilities increase with higher initial vibration levels of nuclei, but they take maximum at a certain level ($v = 9$ for $I = 2 \times 10^{14}\, W/cm^2$; $v = 11$ for $3 \times 10^{14}\, W/cm^2$; $v = 15$ for $4 \times 10^{14}\, W/cm^2$), then decrease with next levels. Those results are obviously illustrated in Fig.2.

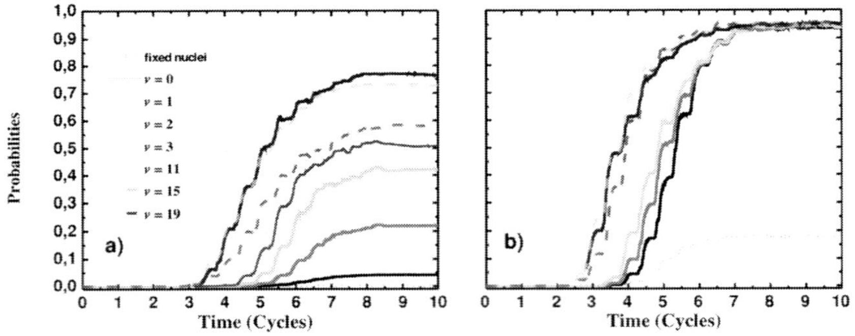

Figure 1. The ionization probabilities of H_2^+ in the field of laser pulses with a wavelength of 800 nm, pulse duration of 26 fs, and peak intensities of (a)$3 \times 10^{14}\, W/cm^2$, (b)$8 \times 10^{14}\, W/cm^2$ for two cases: fixed (dot line) and moving (solid and dashed lines) nuclei.

Since the internuclear distance is greater with increasing nuclear oscillation of the H_2^+, the ionization potential decreases and electron may easily be ionized into the continuum region. To explain the above results of ionization probabilities, the timedependent nuclear separations at various vibration levels of nuclei are also calculated. The results are plotted in Fig. 3. The laser with peaks intensities of $2 \times 10^{14}\, W/cm^2$ and $4 \times 10^{14}\, W/cm^2$, wavelength of 800 nm and pulse duration of 26 fs are used. Initially, the higher vibration level of molecular nuclei is, the

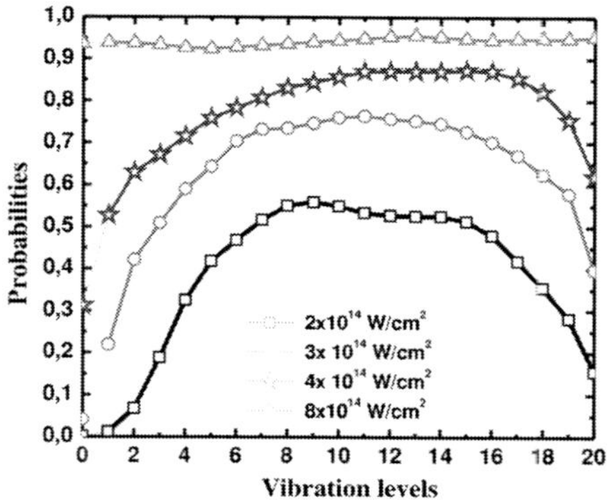

Figure 2. The ionization probabilities of H_2^+ when lasers switch off. Laser pulses with duration of 26 fs, wavelength of 800 nm, and various intensities are used.

greater internuclear separation is (Fig. 3a). During interaction, the nuclear distances are almost unchanged in the some first laser cycles, then rapidly increase in the next cycles (Fig. 3b). The nuclear separation in case of the initial ground state after the laser is switched off is $R_0 = 3\,a.u.$, and is 13% greater than its initial value. The difference of internuclear distances when the laser is turned off between vibration level $v = 0$ and $v = 1$ is $\Delta R_{01} = 2.22\,a.u.$; between $v = 1$ and $v = 2$ is $\Delta R_{12} = 1.39\,a.u.$; $\Delta R_{23} = 1.35\,a.u.$; $\Delta R_{34} = 1.01\,a.u.$; $\Delta R_{45} = 0.64\,a.u.$; $\Delta R_{56} = 0.5\,a.u.$; and others. It is obvious that the differences of separations between the lower vibration states are considerable, but gradually reduced and smaller for higher vibration states. This is the reason causing the above conclusion that the ionization probabilities sharply increase with some first vibration levels, slightly increase with next levels, then almost remained unchanged with higher levels.

While molecular ion H_2^+ interacts with an ultrashort laser pulse of greater intensity ($4\times10^{14}\,W/cm^2$), the internuclear distance rapidly increases even if the nuclei oscillate in ground state ($R_0 = 8.3\,a.u.$) (see in Fig.3a). Therefore, electron is easily ionized, and that leads to a high value of ionization probabilities (31%). The differences of nuclear separations between various vibration levels are $\Delta R_{01} = 1.35\,a.u.$; $\Delta R_{12} = 1.02\,a.u.$; $\Delta R_{23} = 0.83\,a.u.$; $\Delta R_{34} = 0.73\,a.u.$; etc. These values are smaller than that in the case of the lower intensity laser ($2 \times 10^{14}\,W/cm^2$). Thus, when the molecular ion H_2^+ exposed to the laser with higher intensities, the ionization probabilities not only increase more slowly with some first oscillation levels, but also almost is remained with higher vibration levels, compared with the case of lower intensities laser.

The dependences of ionization probabilities on nuclear vibration states may be explained by a combination of the Coulomb potential and the induced-laser potential. In Fig. 4 we plot the total potential energy when the electric field reaches its peak. In the absence of the laser field (bold solid line), the electron is tightly confined in the potential well. Appearance of the laser field leads to the deformation of the potential, and decreases of the potential barrier, thus the tunnel ionization probabilities into continuum region are enhanced. Since the higher nuclear vibration level, the greater internuclear separation, as a consequence the potential well becomes shallower and the ionization rate increases. However, we find from Fig. 1 and Fig. 2 that the ionization probabilities will take maximum at a certain level, then decrease with next levels. This phenomenon can be understood by considering the existence of charge-resonance states of

Journal of Physics: Conference Series **627** (2015) 012014 doi:10.1088/1742-6596/627/1/012014

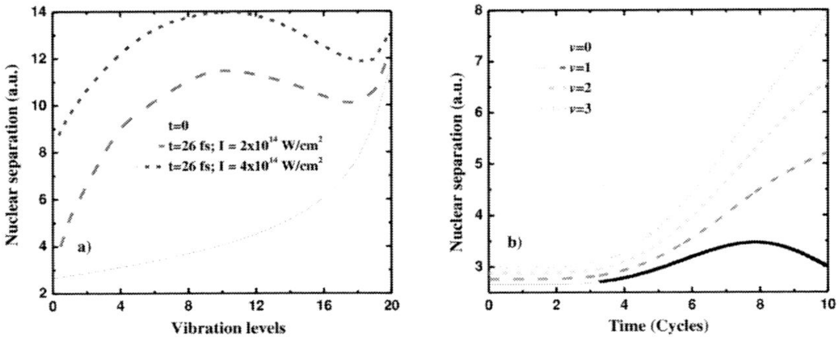

Figure 3. The internuclear separation of H_2^+ at various initial vibration levels before interacting with laser (Fig. a, solid line) and after turning off the laser (Fig. a, dashed line). Fig.b the time-dependent nuclear separation of H_2^+ while interacting with laser with time duration of 26 fs, wavelength of 800 nm, and peak intensity of $2 \times 10^{14}\, W/cm^2$.

odd-charge molecular ions in intense laser field. In [6] authors investigated in detail the process of ionization of diatomic ion H_2^+ in intense lasers as a function of the internuclear separation, and predicted that ionization rate reaches its maximum at critical nuclear distances. In that sense, the electron population of the excited energy level $1\sigma^+$ is more sufficient than that in ground state $1\sigma^-$, thus the dominance of above-barrier ionization leads to the enhanced ionization. If the internuclear separation is greater than the critical value, the above-barrier ionization of excited level does not occur due to decreasing of inner potential well, that leads to a decrease of ionization probability.

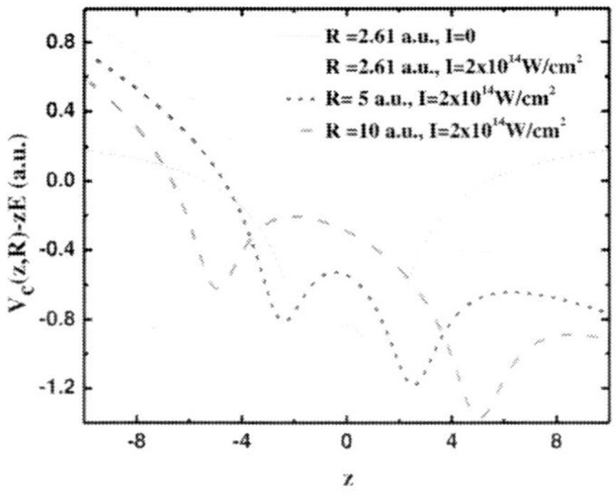

Figure 4. The total potential $V_c(z, R) - zE_0$ for various values of internuclear separations. The corresponding peak intensity of the laser pulse is $2 \times 10^{14}\, W/cm^2$.

The relative velocities between the two nuclei as a function of laser interaction time are exhibited in Figure 5a. Figure 5b presents the dependences of nuclear velocities on vibration levels when the lasers are turned off. We show that when the nuclei oscillate at high vibration states, the relative velocities increase with interaction time. At the low vibration levels, nuclear speeds only increase with some first laser optical cycles, then decrease before the lasers are turned off. Furthermore, in the velocity curves of nuclei, oscillating at low vibration levels, the harmonic properties (appearance of local maximum and minimum) can be seen more clearly. The higher the vibration state is, the more quickly the nuclei move. However, after a certain vibration level, the nuclear relative velocities tend to decrease (Fig. 5b).

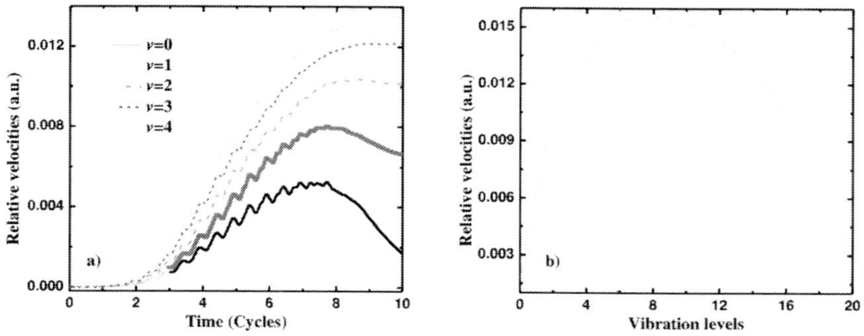

Figure 5. The dependences of relative velocities between the two nuclei on (a) interaction time and (b) vibration levels when the laser is turned off. Using laser pulse with time duration of 26 fs, wavelength of 800 nm, and intensity of $3 \times 10^{14}\ W/cm^2$.

4. Conclusions

By numerically solving the TDSE we obtain the ionization probability of a vibrating H_2^+ exposed to ultrashort intense laser fields at various vibration levels. The more strongly nuclei oscillate, the more ionization probabilities are. However, with higher vibration levels, the ionization probability will take maximum at a certain level and decrease with next levels. The internuclear separations also increase with increasing vibration states, but more slowly with higher levels. The changes of internuclear separations are the basic to explain the obtained dependences of ionization probabilities on initial nuclear vibration levels.

Acknowledgments

This research is funded by Vietnam National Foundation for Science and Technology Development (NAFOSTED) under grant number "103.01-2013.37"

References

[1] Maiman T H 1960 *Nature* **187** 493
[2] Lewenstein M, Balcou Ph, Ivanov M Y, LHuillier A, Corkum P. B. 1994 *Phys. Rev.Lett.* **49** 2117
[3] Zuo T, Chelkowski S, Bandrauk A D 1993 *Phys. Rev. A* **48** 3837
[4] Corkum P B 1993 *Phys. Rev. Lett.* **71** 1994
[5] Bandrauk A D 1993 *Molecules in Laser Fields* (New York: Marcel Dekker)
[6] Zuo T, Bandrauk A D 1995 *Phys. Rev. A* **52** 2511
[7] Keldysh L V 1965 *Sov. Phys. JETP* **20** 1307
[8] Qu W X, Chen Z Y, Xu Z Z, Keitel C H 2001 *Phys. Rev. A* **65** 013402
[9] Guo Y H, He H X, Liu J Y, He G Z 2010 *J. Mol. Struc: THEOCHEM* **947** 119

Journal of Physics: Conference Series **627** (2015) 012014 doi:10.1088/1742-6596/627/1/012014

[10] Bandrauk A D, Chelkowski S and Lu H 2009 *J. Phys. B* **42** 075602
[11] Do T T H, Nguyen N T. 2013 *Journal of Science HCMC UP.* **47** 184
[12] Feit M D, Fleck J A, and Steiger A 1982 *J. Comput. Phys.* **47** 412
[13] Kosloff R, Tal-Ezer H 1986 *Chem. Phys. Lett.* **127** 223

Effective mass of photon in liquid water environment of life and soft maters

Bui Thi Le Quyen[1], **D Ngoc Mai**[2], **N Tri Lan**[2], **and N A Viet**[2]

(1) Thai Binh University of Medical and Pharmacy, 373 Ly Bon, Thai Binh, Viet Nam
(2) Institute of Physics, 10 Dao-tan, Ba-dinh, Hanoi, Viet Nam

Abstract. It is well known that from four fundamental forces, only electro-magnetic force plays clear and important role in our life reality. The spin one exchange boson caring the electro-magnetic interaction is photon. The condition of zero-mass of photon leads to the Coulomb law with long-range interaction behavior. We consider analogy Anderson-Higgs mechanism, in some liquid water environments at a certain temperature, photon can have a finite effective mass, consequence to a Yukawa type potential. This screening potential could leads to sort-range behavior of electro-magnetic effective interaction. We use this mechanism to develop physics models to explain some phenomena in life and soft systems.

1. Introduction

From four fundamental forces, only electro-magnetic force plays clear and important role in our life reality (see figure 1).

According to gauge invariance the exchange boson spin 1 must be massless. In reality, only photon has zero mass, the other exchange particles have a mass, $M_W \approx 80.4GeV$, $M_Z \approx 91.2GeV$; $M_\pi \approx 13510^{-3}GeV$. The problem origin of mass of exchange vector bosons is interesting and important in history of physics.

In the beginning of sixties years of last century, it is known that there are two sources of zero mass bosons: from gauge field theory (because of gauge invariance) which dont correspond to anything to see, and from spontaneous symmetry breaking (SSB) of a continuous symmetry (Nambu-Goldston particles) also are not observed, so both theories had a huge problem. Because of the Goldstone theorem, no observed massless spin 0 boson (Nambu-Goldston particle) means that no spontaneous breaking of a continuous symmetry [7, 8, 9]. The problem also hard in relativistic field theory: spontaneous symmetry breaking implied the existence of massless spineless boson, since no such bosons had been seen, so spontaneous symmetry breaking was rule out, while other models with explicit symmetry breaking were clearly divergent giving infinite results.

Anderson non-relativistically [1], and a lit a bit later Higgs [4, 5] and the others [2, 3] relativistically way gave outstanding answer to that trouble problem: when both gauge symmetry and spontaneous symmetry breaking exist, the Nambu-Goldstone massless mode can combine with the massless gauge field modes to produce a physical massive vector field (exchange boson).

Now a day a type of quasi-particles in condensed mater: plasmon becomes more and more important in modern technology. In practical, we are going from electronics (with the main

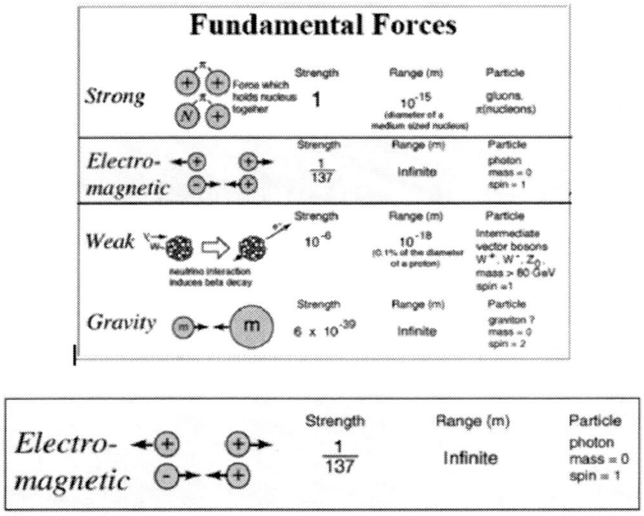

Figure 1. From four fundamental forces, only electro-magnetic force plays clear and important role in our life reality.

particle: electron) to photonics (with the main particle: photon). With photonics, because photon has no rest mass as electron so in one side is good: operating time of photonic devices is speed up and loss energy is less, but in another side is bad because photon interacting time to short so difficult to control and operate them. Plasmonics a new technology with the main particle: plasmon, this trouble is overcame and solved.

We take special attention to the spontaneous breaking of gauge symmetry, giving to plasmon mass, was known in superconductivity [1]. Nambu [7] suggested a similar mechanism could give mass to elementary particles. The condensate of charged pairs (Cooper pair) breaks the gauge invariant of medium. There were found intrinsically massles collective excitations of pairs (NG modes) that restored broken symmetries, and they turned into the plasmon by mixing with the Coulomb field.

In this work we assume that bio liquid water (pH=7.3, salt concentration n=0.9%) is a bio-vacuum with some kind of bio-plasmon. Following the non-relativistic Anderson-Higgs mechanism and Nambu idea [1, 7, 8, 9] on mass problem of plasmon in superconductivity, but instead of photon in a real plasma, we take to investigation mass problem of quasi-photon in bio-plasma (bio water).

We use the unit system with $\hbar = c_0 = 1$ where c_0 is speed of photon in vacuum.

2. Model for Polymer chains

Define end-to-end vector R (see the figure 2) [6]

$$R = r_N - r_0 = \sum_{i=1}^{N} a_i \tag{1}$$

We assume a simple model for zip-unzip transformation of polymer like Cooper pair formation mechanism in superconductivity with two effective charges Q in a bio-water vacuum. This mechanism is well presented in language of spontaneously symmetry broken with a Mexican hat

 IOP Publishing
doi:10.1088/1742-6596/627/1/012015

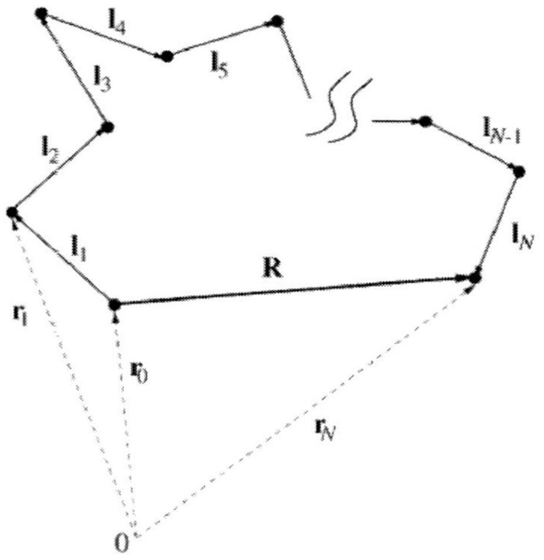

Figure 2. A simple discription of a polymer chain. The filled circles denote atoms in those positions, the bond length is a, R is the end-to-end vector.

like potential.
According to gauge invariance the exchange boson spin 1 must be massless. In the case of massless bio-photon (exchange boson), electro-magnetic effective potential between two charge Q is

$$W_0(r) = \frac{Q^2}{\varepsilon * r} \tag{2}$$

where $\varepsilon*$ is the dielectric constant of bio-water, r is distance between two charges. This process can be expressed by a Feynman diagram in figure 3.

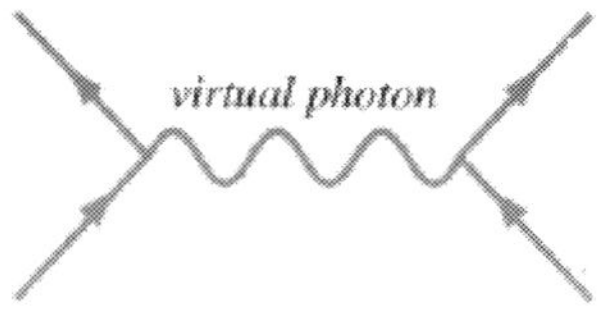

Figure 3. Feynman diagram for the electromagnetic interaction between two charges.

By mean of Anderson-Higgs mechanism with bio-plasmon, bio-photon can be massive by spontaneously broken some symmetry (denoted as long-sort symmetry) mechanism (see figure 4).
Using the Yukawa idea of massive exchange pi-meson in strong interaction, we assume the Yukawa like potential for electro-magnetic effective potential in bio-water

$$W(r) = \frac{Q^2}{\varepsilon * r} \cdot \exp(-mr) \tag{3}$$

where $\varepsilon*$ is the dielectric constant of bio-water, m is effective of bio-photon (exchange boson).

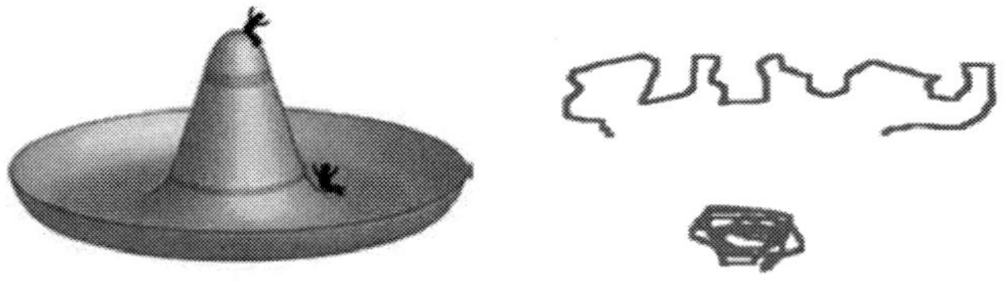

Figure 4. Spontaneously symmetry broken problem with a Mexican hat like potential.

We take $m = 1/R$ as a parameter of our model.
For the case of ideal (or Gaussian) unzipping chains

$$R_0 = \sqrt{\langle R^2 \rangle} = aN^{1/2} \tag{4}$$

where a is the bond length, N is total number of sites in the chain. In the case of a long chain $N \to \infty$, $R_0 \to \infty$, so $m = 1/R_0 \to 0$, bio-photon is nearly massless.
If the chain go to its zipping state with $R = b << R_0$ then $m = 1/b$, bio-photon now became massive.
A typical zip-unzip phase transition of polymers is presented in Figure 5.

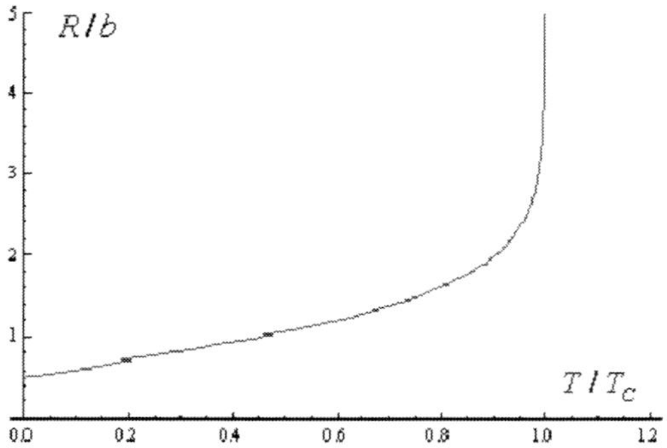

Figure 5. A typical zip-unzip phase transition of.

3. Ginzburg-Landau formalism

We express the phase transition between zipping and unzipping states of a polymer by Ginzburg-Landau formalism.
Introducing the free energy density $f(\psi, T)$ which is a function of the real local order parameter ψ and temperature T in the form

$$f(\psi, T) = f_0(T) + \alpha (T - T_C) \psi^2 + \frac{1}{2}\beta\psi^4 \tag{5}$$

where α and β are some constants, T_c is the critical temperature.

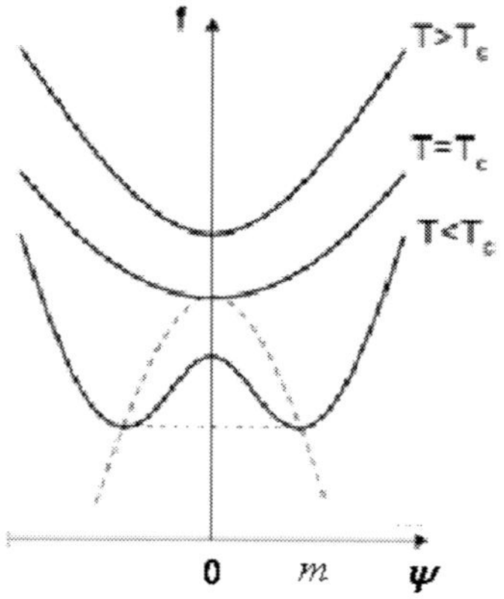

Figure 6. Free energy density f as a function of ψ. Below T_c new minima at nonzero $\pm\psi_0 = \pm m$ develop (the dashed curve denotes $m(T)$).

The value ψ_0 corresponding a minimum are

$$\psi_0 = 0 \qquad \text{for } T > T_C \tag{6}$$

$$\psi_0 = \left(\frac{\alpha}{\beta}\right)^{1/2} (T_C - T)^{1/2} \qquad \text{for } T > T_C \tag{7}$$

Define here $\psi_0 = m = \frac{1}{b}$ is bio-photon mass. The values of bio-photon mass ratio m/m_0 where $m_0 = (\alpha/\beta)^{1/2} {T_C}^{1/2}$ versus temperature are presented in the figure 7.

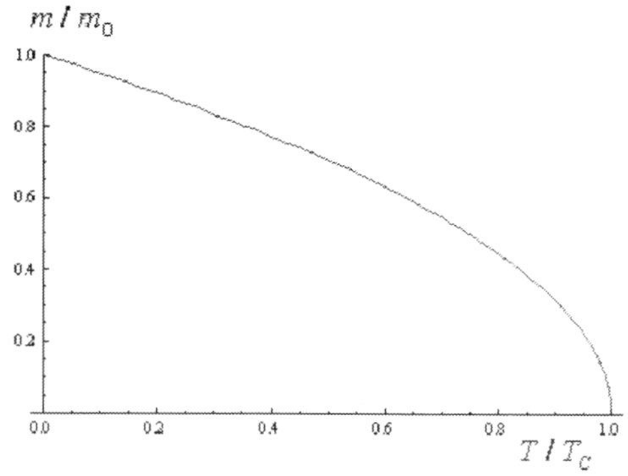

Figure 7. The values of bio-photon mass ratio m/m_0 versus temperature.

Evaluating f at ψ_0 gives

$$\psi_0 = 0 \qquad \text{for } T > T_C \tag{8}$$

$$f = f_0(T) - \frac{\alpha^2}{2\beta}(T_C - T)^2 \qquad \text{for } T > T_C \tag{9}$$

showing the lowering free energy under transition temperature.

Journal of Physics: Conference Series **627** (2015) 012015 doi:10.1088/1742-6596/627/1/012015

Note that f deviates from f_0 quadractically, this will yield a jump discontinuity in the specific heat $\Delta C = (\alpha/\beta)T_C = m_0{}^2$ at $T = T_C$. So that measure the jump discontinuity in the specific heat at critical temperature by experiments could give us the maximum mass value of bio-photon.

4. Discussion

In this work we consider a new type of quasi-particle for bio systems in bio-water vacuum and called it bio-photon. In analogy Anderson-Higgs mechanism, in liquid water environments under critical temperature, bio-photon can have a finite effective mass, consequence to a Yukawa type potential. This screening potential could leads to sort-range behavior of electro-magnetic effective interaction like Cooper pairs formulation in superconductivity. Based on Ginzburg-Landau formalism, we proposed a simple physics model to describe zip-unzip phase transition of polymers. This new model might be useful in investigation other similar bio systems.

References

[1] Andeson P.W (1963) *Phys. Rev.* **130** 439.
[2] Englert F and Brout R (1964) *Phys. Rev. Lett.* **13** 321.
[3] Guralnik G.S, Hagen C.R and Kibble T.W.B (1964) *Phys. Rev. Lett.* **13**, 585.
[4] Higgs P.W (1964) *Phys. Lett.* **12** 132.
[5] Higgs P.W (1964) *Phys. Rev. Lett.* **13** 508.
[6] Likos C N (2001) *Physics Reports.* **348** 267.
[7] Nambu Y (1960) *Phys. Rev.* **117** 648.
[8] Nambu Y (1960) *Phys. Rev. Lett.* **4** 380.
[9] Nambu Y and Jona Lasino G (1961) *Phys. Rev.* **122** 345.

Journal of Physics: Conference Series **627** (2015) 012016 doi:10.1088/1742-6596/627/1/012016

The statistical distribution of (q, R)-deformed crystal lattice vibration for generic atomic string

Nguyen Thi Ha Loan[1], **Nguyen Anh Sang**[2], **Do Thi Thu Thuy**[3]

[1] Ha Noi Pedagogical University No2, Xuan Hoa, Phuc Yen, Vinh Phuc, Vietnam
[2] Vocational Training and Comprehensive Technical Center of Phu Tho, Phu Tho, Vietnam
[3] Cam Pha Industrial College, Cam Thuy, Cam Pha, Quang Ninh, Vietnam

E-mail: hloansp2@gmail.com, anhsangk13@gmail.com, thuthuy5585@gmail.com

Abstract. The deformed oscillators formalism plays a useful role in the study of physical models. Specifically, many variants of deformed oscillators have been used to study of the quantum optics, condensed matter physics, etc ... In this paper we construct a (q, R)-deformed crystal lattice vibration for generic atomic string and calculate statistical distribution of this vibration. This is a fundament for study of crystal lattice vibration, which would give a base for a new quantization procedure.

1. Introduction

The study of q-quantum algebras and their oscillator representation [1, 2, 3, 4, 5, 6] is a direction of actual character in theoretical physics. They find application in many problems such as quantum inverse scattering theory, exactly solvable model in statistical mechanics, rational conformal field theory, two-demensional field theory with fractional statistics, etc. The algebraic structure of quantum group can be formally described as a deformation, depending on one or more parameters of the classical Lie algebras [7, 8]. In the special limiting cases of these parameters the quantum algebra reduce to the ordinary Lie algebras. Especially the oscillator representation of quantum algebras has proved to be powerful for the study of the quantum optics, condensed matter physics, etc.

The R-deformed Heisenberg algebra, which is the deformation involving the reflection operator $R(R^2 = 1)$, was introduced by Vasiliev [9] in the context of the higher spin algebras. The (q, R)-deformed Heisenberg algebra is combined the q-deformed algebra with the R-deformed, which may be described as a deformation, depending on q-parameter and reflection operator $R(R^2 = 1)$ of the ordinary Lie algebra.

The above mentioned arguments have encouraged us to consider the possibility for the application of the quantum group concepts to the problem in the theory of solid physics.

Our aim here is to construct a (q, R)-deformed crystal lattice vibration for generic atomic string, consider a statistical distribution in view of this vibration.

2. The (q, R)-deformed crystal lattice vibration for generic atomic string

In this section, we are using the result in our previous paper [10] to construct a (q, R)-deformed crystal lattice vibration for generic atomic string.

Journal of Physics: Conference Series **627** (2015) 012016 doi:10.1088/1742-6596/627/1/012016

The Hamiltonian of the crystal lattice vibration for generic atomic string can be expressed following form:

$$H = \sum_{k \in BZ_1} \left(\frac{p_k^2}{2m} + \frac{1}{2} m\omega^2 (k)\, x_k^2 \right), \tag{1}$$

where k is wave vector and $\sum_{k \in BZ_1}$ is a sum on k, which is one in first Brillouin zone, the x_k is extended coordinate of atom in order k and the p_k is correspondent momentum of generic atomic string.

The creation and annihilation operators, which are to correspond to waved vector k, are form

$$\begin{aligned} b_k^{\dagger} &= \frac{1}{\sqrt{2\hbar\omega(k)}} \left(\sqrt{m}\omega (k)\, x_k - \frac{i}{\sqrt{m}} p_k \right) \\ b_k &= \frac{1}{\sqrt{2\hbar\omega(k)}} \left(\sqrt{m}\omega (k)\, x_k + \frac{i}{\sqrt{m}} p_k \right). \end{aligned} \tag{2}$$

The coordinate and momentum operators can be expressed in creation and annihilation operators $b^{\dagger}, b$ as:

$$\begin{aligned} x_k &= \sqrt{\frac{\hbar}{2m\omega(k)}} \left(b_k + b_k^{\dagger} \right) \\ p_k &= -i\sqrt{\frac{\hbar m\omega(k)}{2}} \left(b_k - b_k^{\dagger} \right). \end{aligned} \tag{3}$$

Using equations (1) and (3), one obtains the Hamiltonian of the crystal lattice vibration for generic atomic string in following form:

$$H = \sum_{k \in BZ_1} \frac{\hbar\omega (k)}{2} \left(b_k b_k^{\dagger} + b_k^{\dagger} b_k \right). \tag{4}$$

The creation (annihilation) operator $b_k^{\dagger}$ (b_k) of the equation (4) satisfy the commutation relation:

$$b_k b_k^{\dagger} - b_k^{\dagger} b_k = 1. \tag{5}$$

The (q, R)-deformed crystal lattice vibration can be considered as nonlinear crystal lattice vibration which its frequency depending on amplitude. Using the analogical method in the work [11] we calculated a frequency depending on amplitude, such as:

$$W_{q\nu} (k) = \left\{ \frac{\cosh \left(\hbar bb^{\dagger} \right)}{\sinh \hbar} + \frac{\hbar e^{\hbar bb^{\dagger}}}{e^{\hbar} - 1} \nu \right\} \omega (k).$$

Instead of amplitude dependence of frequency the Hamiltonian could be rewritten in the form of standard Hamiltonian as:

$$H = \sum_{k \in BZ_1} \frac{\hbar\omega (k)}{2} \left(a_k a_k^{\dagger} + a_k^{\dagger} a_k \right), \tag{6}$$

where the creation (annihilation) operator $a_k^{\dagger}$ (a_k) and the number operators N_k of the (q, R)-deformed crystal lattice vibration satisfy the commutation relations:

$$\begin{aligned} a_k a_{k'}^{\dagger} - q a_{k'}^{\dagger} a_k &= q^{-N_k} \delta_{k,k'} + \nu R \\ \left[a_k, a_{k'} \right] &= 0 \\ \left[N_k, a_{k'} \right] &= -a_k \delta_{k,k'} \\ \left[N_k, a_{k'}^{\dagger} \right] &= a_{k'}^{\dagger} \delta_{k,k'}, \end{aligned} \tag{7}$$

with R is reflection operator satisfying the following commutation relations:

$$Ra_k^\dagger + a_k^\dagger R = Ra_k + a_k R = 0, \tag{8}$$

and frequence is now not depending on the amplitude.
The (q, R)-deformed state vectors in Fock space are

$$|n_k\rangle_{q\nu} = \frac{\left(a_k^\dagger\right)^{n_k}}{\sqrt{[n_k]_{q\nu}!}}|0\rangle_{q\nu}, \tag{9}$$

where $|0\rangle_{q\nu}$ is the ground state satisfying the following conditions:

$$\begin{aligned} a_k|0\rangle_{q\nu} &= 0 \\ \langle 0 \mid 0\rangle_{q\nu} &= 1 \\ N_k|0\rangle_{q\nu} &= 0 \\ R|0\rangle_{q\nu} &= r|0\rangle_{q\nu} \\ r &= \pm 1, \end{aligned} \tag{10}$$

using the notations:

$$\begin{aligned} [n_k]_q &= \tfrac{q^{n_k} - q^{-n_k}}{q - q^{-1}} \\ [n_k]_{q\nu} &= [n_k]_q + \tfrac{q^{n_k} - (-1)^{n_k}}{q-1}\nu \\ [n_k]_{q\nu}! &= [n_k]_{q\nu}.[n_k - 1]_{q\nu}\dots[1]_{q\nu}. \end{aligned} \tag{11}$$

Let $|n_k\rangle_{q\nu}$ be the eigenstates of the oscillator number operator

$$N_k|n_k\rangle_{q\nu} = n_k|n_k\rangle_{q\nu}. \tag{12}$$

From (7) one sees that: The action of operators $a_k^\dagger$ and a_k on the eigenstates $|n_k\rangle_{q\nu}$ can be chosen as

$$\begin{aligned} a_k^\dagger|n_k\rangle_{q\nu} &= \sqrt{[n_k+1]_{q\nu}}|n_k+1\rangle_{q\nu} \\ a_k|n_k\rangle_{q\nu} &= \sqrt{[n_k]_{q\nu}}|n_k-1\rangle_{q\nu}. \end{aligned} \tag{13}$$

The energy spectrum of (q, R)-deformed crystal lattice vibration of generic atomic string is given by

$$\begin{aligned} H|n_k\rangle_{q\nu} &= E_{n_k}|n_k\rangle_{q\nu} \\ \sum_{k\in BZ_1} \tfrac{\hbar\omega(k)}{2}\left(a_k a_k^\dagger + a_k^\dagger a_k\right)\tfrac{\left(a_k^\dagger\right)^{n_k}}{\sqrt{[n_k]_{q\nu}!}}|0\rangle_{q\nu} &= E_{n_k}|n_k\rangle_{q\nu}, \end{aligned}$$

where

$$E_{n_k} = \sum_{k\in BZ_1} \frac{\hbar\omega(k)}{2}\left\{[n_k+1]_{q\nu} + [n_k]_{q\nu}\right\}. \tag{14}$$

Hence, the energy spectrum E_{n_k} of the (q, R)-deformed crystal lattice vibration for generic atomic string have been depended on deformed parameters in the formulation (14).

Journal of Physics: Conference Series **627** (2015) 012016 doi:10.1088/1742-6596/627/1/012016

3. The Green function for (q, R)-deformed crystal lattice vibration for generic atomic string

The Green function for (q, R)-deformed crystal lattice vibration for generic atomic string is defined as the statistical distribution of $\left(a_k^\dagger a_k\right)$

$$\left\langle a_k^\dagger a_k \right\rangle = \frac{1}{Z}\mathrm{Tr}\left(e^{-N_k\beta\omega(k)}a_k^\dagger a_k\right), \tag{15}$$

where Z is the partition function

$$\begin{aligned} Z &= \mathrm{Tr}\left(e^{-N_k\beta\omega(k)}\right) \\ &= \sum_{n_k}^{\infty}{}_{q\nu}\left\langle n_k\right|e^{-N_k\beta\omega(k)}\left|n_k\right\rangle_{q\nu}, \end{aligned} \tag{16}$$

which could be used to determine the thermodynamic properties of the system, $\beta = \frac{1}{kT}$, $\omega(k)$ is one particle-oscillator energy. The trace most be taken over a complete set of states.
The calculations based on the equation (9) give the following results:

$$\left\langle a_k^\dagger a_k \right\rangle = \frac{\left(e^{\beta\omega}-1\right)}{\left(e^{\beta\omega}-q\right)}\left\{\frac{1}{e^{\beta\omega}-q^{-1}} + \frac{\nu}{e^{\beta\omega}+1}\right\}.$$

In the limit $q \to 1$ and $\nu \to 0$, this gives

$$\left\langle a_k^\dagger a_k \right\rangle = \frac{1}{e^{\beta\omega}-1},$$

which is just the usual Bose-Einstein formulae.

4. Conclusion

In this work, we present the results of our primary calculations which have been performed on the model of (q, R)-deformed crystal lattice vibration that can be realized by several symmetric and asymmetric potentials in solid systems such as Rogers-Szegö and Stieltjes-Wigert polynomials. The most important properties of potentials in the solid systems are the finite number of energy levels and the non-equal steps between energy levels that would not be described by the model of harmonic oscillators, however, that are possible to transform into the model of (q, R)-deformed harmonic oscillators by one-to-one mapping. Our results provide the initial powerful tools to study Einstein and Debye solid models to obtain thermodynamic quantities such as Einstein and Debye temperatures, specific heat, thermal conductivity, electrical conductivity and resistivity with $q \neq 1$ that will be calculated in next works.
Our results have shown that it is possible to apply q-deformation in condensed matter physics and to investigate the causes of other factors within materials in definition of q-deformation.

References

[1] Drinfeld V G 1985 *Sov. Math. Dokl* **32** 254
[2] Drinfeld V G 1986 *Proceedings of the International Congress of Mathematicians, Berkeley, CA, USA* 793
[3] Jimbo M 1985 *Lett. Math. Phys.* **10** 63
[4] Kinani E H El and Zakhnri M 1995 *Phys. Lett.* B **357** 105
[5] Leinaas M and Myrhein J 1988 *Phys. Rev.* B **37** 9286
[6] Woronowicz S L 1987 *Comm. Math. Phys.* **111** 613
[7] Aizawa N and Sato R 1991 *Phys. Lett.* B **265** 185
[8] Brink L, Hansson T H and Vasiliev M A 1992 *Phys. Lett.* B **286** 109
[9] Vasiliev M A 1989 *Pis'ma JETP* **50** 344; 1991 *Int. Mod.* A **6** 1115
[10] Loan N T H and Ha N H 2003 *Comm. in Phys.* **13 No4** 240
[11] Man'ko V I, Marmo G, Solimeno S and Zaccaria F 1993 *Int. J. Mod. Phys.* A **8** 3577

Journal of Physics: Conference Series **627** (2015) 012017 doi:10.1088/1742-6596/627/1/012017

Trapping Cold Atoms Using Surface Plasmons with Phase Singularities Generated by Evanescent Bessel Beams

Nguyen Thi Phuong Lan[1], **Dao Thi Thuy Nga**[2] **and Nguyen Ai Viet**[3]

[1]Hanoi Pedagogical University No.2, Nguyen Van Linh Street, Phuc Yen Town, Vinh Phuc Province, Vietnam

[2]Han Thuyen high school, Bac Ninh City, Bac Ninh Province, Vietnam

[3]Institute of Physics, Vietnam Academy of Science and Technology, 10 Dao Tan, Thu Le, Ba Dinh, Ha Noi, Vietnam

E-mail: `lanntphuong@gmail.com`

Abstract. When a laser Bessel beam light is totally reflected internally at the planar surface of a dielectric on which a thin film has been deposited, a surface plasmons with phase singularities can be produced. This two-dimensional plasmons lead to an attractive potential with enhancement. Using this effect, we construct a model for optical trapping cold atoms. Efficiency of the trap will be presented by an ellipse with average radius r in the (x, y) directions and the distance d from planar surface in the z-direction.

1. Introduction

Trapping methods of neutral atoms are interest in both basics side as pure quantum effects and in application side in many modern technologies such as nano, bio, electro, quantum optic. Cold trapped radioactive atoms can be used in fundamental symmetry experiments, including the experiments on nuclear decay, atomic parity non-conservation and the search for parity and time-reserve violating electric dipole moment [1]. Trapping of ultracold atoms also gives us an opportunity to study the collision processes in cold atomic samples. Trapped cold atoms can be used in the formation of cold molecules [2]. Since the cold molecule trap lifetime was approximately half a second, so the production of cold molecules opens new ways of research in molecular spectroscopy. It has been shown theoretically and experimentally that neutral atoms can be captured by an optical trap [3], based on the use of an evanescent wave occurred around a quasi-one-dimensional micro-nano structure, such as optical fiber [4], thin silica fiber [5], carbon nanotube [6-8], nano pilar [9]. In those optical traps, the effective optical attractive potential created by light is inversely proportional to the radius of fiber. The condition for existence of the trap is that the fiber diameter is about two times smaller than the light wave length. In the last decade surface plasmon (SP) and surface plasmon polariton (SPP), the collective oscillations of conduction electrons and their coupled with photon at the metal/dielectric interface, receive much attention in physics, chemistry, biology and technology communities. Due to their UV-vis-wavelength and local field enhancement effects, surface plasmons and surface plasmon polaritons found vast application in many areas, near field microscopy, Forster resonance energy transfer (FRET), surface enhanced Raman spectroscopy (SERS), nonlinear optics, bio- and nano sensors

[10-15]. In this work we study the optical trap using the surface plasmon effect. Plasmon and plasmon polaritons theories always are difficult to understand for numerous experimental physicists and technicians, a simple model will be useful and needed. In this work, using a laser Bessel beam light and the Kretschmann configuration of excitation surface plasmons, we propose a simple model of optical trap. In this model the effective volume of the trap will be presented in a cylindrical form with the average radius and height of first bound state. This simple picture could be useful for imagine and design idea of optical plasmon could atom trap.

2. Optical plasmon trap and conditions for surface excitation plasmon

We use the Kretschmann configuration of excitation surface plasmons for optical surface plasmon trap design. The Kretschmann configuration is presented in the figure 1, consisting a metallic thin film M deposited on a planar dielectric substrate denoted as $(x, y, z = 0)$ plane, laser source L, and detector D. A strong laser beam in the $(x, y = 0, z)$ plane produced by the source D

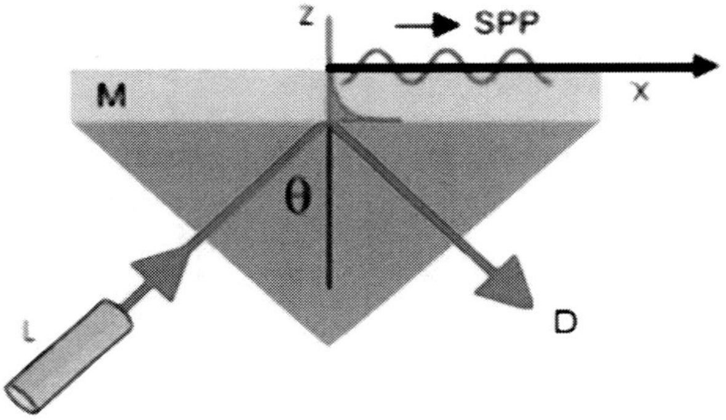

Figure 1. Kretschmann configuration of excitation surface plasmons. M: thin metal layer, L: laser beam source, D: detector.

with frequency ω and radius R_0 is incident at angle θ to surface metallic thin film. Laser beam is reflected at the interface $z = 0$ between the dielectric substrate and the metallic thin film. Consequently a surface plasmon evanescent mode is produced at the vacuum/metal interface. This evanescent mode creates an effective attractive potential, leads to a bound state of a cold atom at the vacuum/metal interface. We can use that effect as working principle of optical plasmon trap for cold atom. The matching conditions for excitation surface plasmon are
i) wave vectors of photon γ and surface plasmon in the x-direction are coincided $k_{photon,x} = k_{Sp,x}$,
ii) incident and the total reflection angles θ_0 at are coincided $\theta = \theta_0$,
iii) metal thickness smaler than skin depth. These surface plasmon excitation requirements are presented in the figure 2.
Denote ε_0 and ε_m are the dielectric constants of dielectric and metal, c is the speed of light, the dispersion laws are

$$\omega = ck, \tag{1}$$

for photon in vacuum or air, and

$$\omega = ck\sqrt{(\varepsilon_m + 1)/\varepsilon_m}, \tag{2}$$

for surface plasmon. The total reflection angle satisfies the relation

$$\sin(\theta_0) = \frac{ck}{\omega\sqrt{\varepsilon_0}} \quad . \tag{3}$$

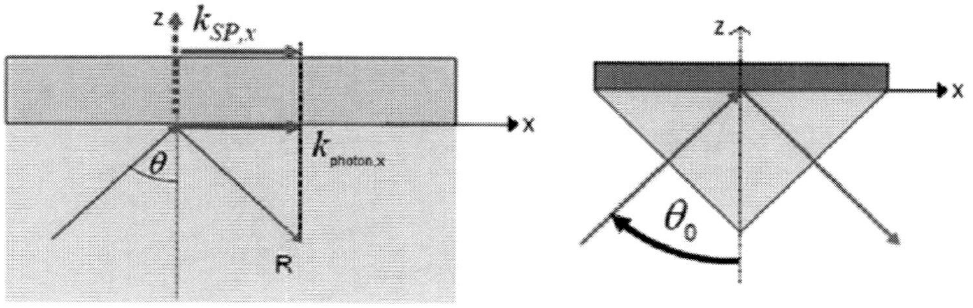

Figure 2. Matching conditions for excitation surface plasmon.

From the condition of equal wave vectors of photon γ at total reflection angle and surface plasmon we have the resonance condition for surface plasmon excitations

$$c\sqrt{(\varepsilon_m + 1)/\varepsilon_m} = c/(\sqrt{\varepsilon_0}\sin\theta_0) = \omega_0/k_0, \tag{4}$$

where ω_0 and k_0 are the resonant frequency and wave vector (see the figure 3).

3. Theory of optical plasmon trap model

Consider a Bessel laser beam is totally reflected at the planar dielectric/metal interface and consequently a surface plasmon evanescent mode is produced at the vacuum/metal interface. The envelop function of a Bessel laser beam order l traveling along z-direction and having electrical vector in the y-direction can written as [15]

$$F_{k_z l}(r, z) = CI_B e^{ik_z z}(z/z_0)^{l+1/2}\exp(-z^2/z_0^2)J_l(r/R), \tag{5}$$

where C is a factor, I_B is the intensity of Bessel laser beam, $r = \sqrt{x^2+y^2}$ is radial vector (parallel to the interface plane), R is effective radius of Bessel laser beam, z_0 is the typical ring spacing of Bessel laser beam, $J_l(r/R)$ is the Bessel function of order l.
Consider light as a vector, when incident angle of Bessel laser beam is θ, a projective operator $P(\theta)$ for separation the out-of plane z direction and in-plane (x, y) directions can be defined by

$$P(\theta)F(\overrightarrow{r}, z) = F_z(\theta) + F_r(\theta). \tag{6}$$

Assuming Bessel laser beam is in (y0z) plane, and the incident angle is θ relative to the z-direction, the projections can be expressed by transforms

$$x \to x/\cos\theta, \quad y \to y, \quad z \to z\cos\theta. \tag{7}$$

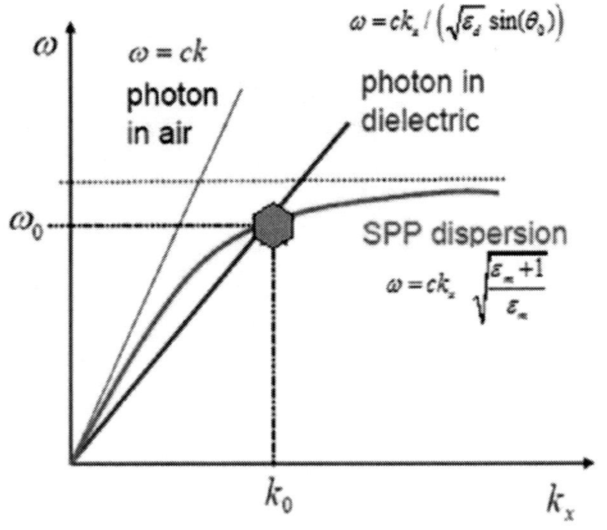

Figure 3. The resonant frequency ω_0 and wave vector k_0 for surface plasmon polariton SPP.

Apply the boundary conditions for electric and magnetic field of Bessel laser beam at the interfaces, we can realize that an evanescence light is exponential decay with distance z in vacuum $\exp(-z/\Lambda)$ where Λ is characteristic decay length of the evanescent field, and provides a two-dimensional surface plasmons with phase singularities and attractive enhancements [15]. The effective attractive optical potential in the z-direction introduced by evanescent effect can be written as

$$V_l(z;\theta) = -gI_B(z\cos\theta/z_0)^{l+1/2}\exp(-z\cos\theta/\Lambda)\cdot\exp(-z^2\cos^2\theta/z_0^2), \tag{8}$$

where g is effective coupling constant for interaction between evanescent wave and cold atom [3-9]. The effective attractive optical potential in the interface (x,y) plane introduced by evanescent effect is

$$U_l(x,y;\theta) = -gI_BJ_l\left(\sqrt{(x^2/\cos^2\theta)+y^2}/R\right). \tag{9}$$

Matching conditions for surface plasmon excitation require that the incident angle must be closed to total reflection angle $\theta = \theta_0$. For many metals, example Au $\theta_0 \sim 41-43^\circ \approx \pi/4$ so a good approximate to put in the expressions of effective potentials (8-9) that $\cos\theta \approx 1/\sqrt{2}$. The effective attractive optical potential in the z-direction introduced by evanescent effect now has a form

$$V_l(z) = -gI_B(z/z_0\sqrt{2})^{l+1/2}\exp(-z/\Lambda\sqrt{2})\cdot\exp(-z^2/2z_0^2), \tag{10}$$

The effective attractive optical potential in the interface (x,y) plane introduced by evanescent effect is

$$U_l(x,y) = -gI_BJ_l\left(\sqrt{2x^2+y^2}/R\right). \tag{11}$$

For the S state ($l=0$), we have

$$V_0(z) = -gI_B(z/z_0\sqrt{2})^{1/2}\exp(-z/\Lambda\sqrt{2})\cdot\exp(-z^2/2z_0^2), \tag{12}$$

Journal of Physics: Conference Series **627** (2015) 012017 doi:10.1088/1742-6596/627/1/012017

$$U_0(r) = -gI_B J_0\left(\sqrt{2x^2+y^2}/R\right). \tag{13}$$

Define $z' = z/z_0\sqrt{2}$ and $t = z_0/\Lambda$. The values of effective potentials V_0 in z-direction as a function of z' and t, and U_0 in (x, y)-plane for the S state $l = 0$ are presented in the figure 4.

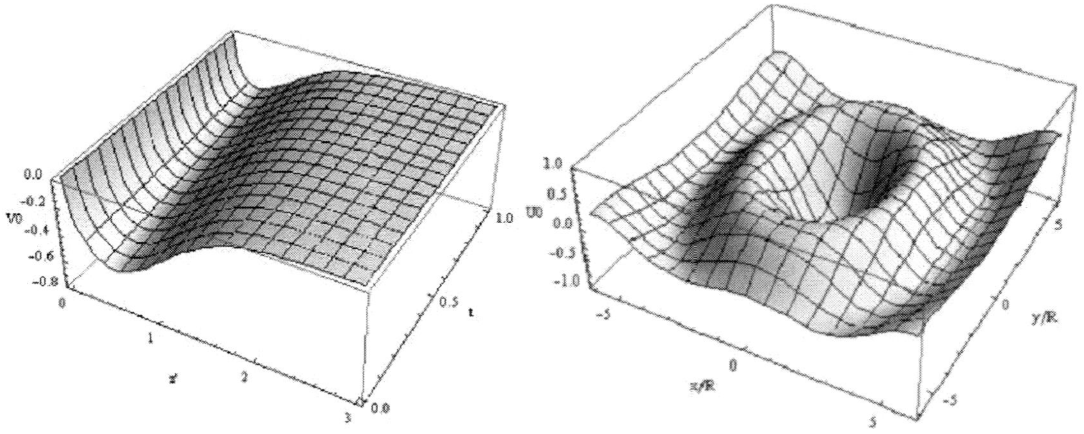

Figure 4. The values of effective potentials V_0 in z-direction as a function of z' and t, and U_0 in (x, y)-plane for the S state $l = 0$.

For the P state ($l = 1$), we have

$$V_1(z) = -gI_B(z/z_0\sqrt{2})^{3/2}\exp(-z/\Lambda\sqrt{2})\cdot\exp(-z^2/2z_0^2), \tag{14}$$

$$U_1(r) = -gI_B J_1\left(\sqrt{2x^2+y^2}/R\right), \tag{15}$$

The values of effective potentials V_1 in z-direction as a function of z' and t, and U_1 in (x, y)-plane for the P state $l = 1$ are presented in the figure 5.

Note that the trapping efficiency is proportional to the intensity of Bessel laser beam and effective coupling constant for interaction between evanescent wave and cold atom. Between the two case of $l = 0$ and $l = 1$, the S state give more trapping efficiency.

4. Simple pictorial model of optical plasmon trap

Plasmon and plasmon polariton theories always are difficult to understand for numerous experimental physicists and technicians, so in this part, we develop a simple pictorial model optical plasmon trap cold atom. Using the well known series expansion of Bessel function

$$J_0(x) \approx 1 - \frac{x^2}{4} + \frac{x^4}{64} + \ldots, \tag{16}$$

and neglecting a constant, we can approximate the effective potential in the z-direction as

$$U_0(x, y) \approx \frac{gI_B}{4R^2}(2x^2 + y^2), \tag{17}$$

The Hamiltonian for motion in (x, y)-directions is

$$H_{Cin}(x, y) = \frac{p_x^2 + p_y^2}{2m} + \frac{gI_B}{4R^2}(2x^2 + y^2), \tag{18}$$

 IOP Publishing
doi:10.1088/1742-6596/627/1/012017

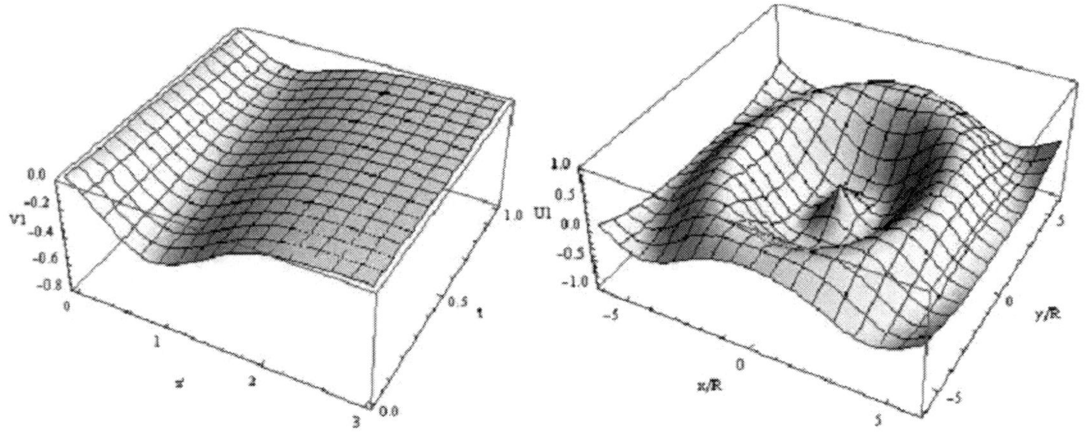

Figure 5. The values of effective potentials V_1 in z-direction as a function of z' and t, and U_1 in (x, y)-plane for the P state $l = 1$.

where m is the mass of cold atom. This Hamiltonian has the form of the Hamiltonian for harmonic oscillator, is easy to solve and gives bound state energy spectrum

$$E_m = \frac{\omega^*}{2}(2n+1), n = 0, 1, 2, ... \tag{19}$$

where ω^* is the characteristic frequency of the optical plasmon trap

$$\omega^* = \left(1 + \frac{1}{\sqrt{2}}\right)\sqrt{\frac{gI_B}{mR^2}} \quad . \tag{20}$$

The energy of ground bound state is

$$E_{\gamma 0} = \frac{3}{2}\omega^* = \frac{3}{2}\left(1 + \frac{1}{\sqrt{2}}\right)\sqrt{\frac{gI_B}{mR^2}} \quad . \tag{21}$$

The trapping effect by evanescent wave in (x, y)-directions can be modeled as an ellipse disc expressed by the equation

$$T_{in}(x, y) = \frac{gI_B}{4R^2}(2x^2 + y^2) \quad . \tag{22}$$

In the general case, when the incident angle of Bessel laser beam equals to the total reflection angle θ_0 we have following expressions for energy of ground bound state

$$E_{r0} = \frac{3}{2\sqrt{2}}\left(1 + \frac{1}{\cos\theta_0}\right)\sqrt{\frac{gI_B}{mR^2}}, \tag{23}$$

and for ellipse disc

$$T_{in}(x, y) = \frac{gI_B}{4R^2}\left(\frac{x^2}{\cos^2\theta_0} + y^2\right). \tag{24}$$

From figure 4, we can see that the effective radius of the effective attractive optical potential in the (x, y) plane introduced by evanescent effect is approximate $R_D \approx 3.8R$.

The t-dependence minimum line $z'_{\rm min}$ of the effective attractive optical plasmon potential in the z-direction introduced by evanescent effect is defined by equation

$$z'_{\rm min} = \left(-t + \sqrt{4+t^2}\right)/4, \qquad (25)$$

and presented in the figure 6. With this $z'_{\rm min}$ later we define the distance d of our trap model

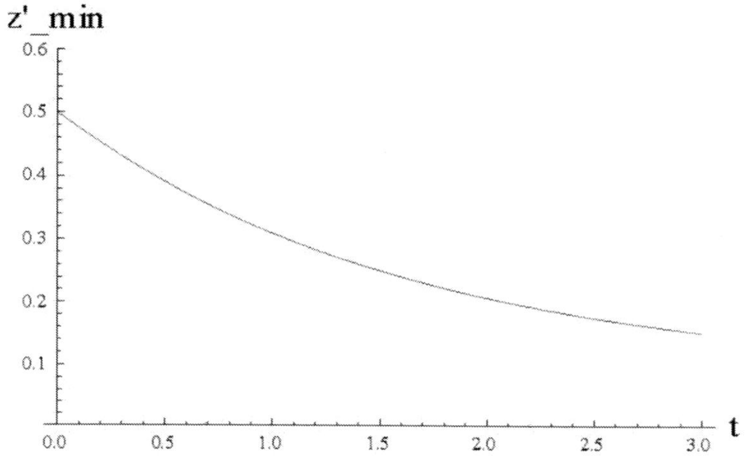

Figure 6. The t-dependence minimum line of the effective attractive optical plasmon potential in the z-direction.

Finally, we suppose a simple pictorial model for the effective attractive optical plasmon potential introduced by evanescent effect. The trap is modeled by an elliptic cylinder as presented in the figure 7. This ellipse expressed by equation (24) with average radius $R_D \approx 3.8R$, and lies above (x, y)-plane a height $d = z'_{\rm min} z_0 \sqrt{2}$, where $z'_{\rm min}$expressed by equation (25) at incident point of the Bessel laser beam. The strength of the trap is gI_B. This simple pictorial model of the optical plasmon trap might be useful for imagination, set up experiment, and technical design, etc.

5. Discussion

The main results of this work is simple theory describing interaction a Bessel laser beam with surface plasmon in thin film metal. Based on detail numerical investigation, we suppose a simple pictorial model for the effective attractive optical plasmon potential introduced by evanescent effect. The trap is modeled by an elliptic cylinder as presented in the figure 7. This ellipse expressed by equation (24) with average radius $R_D \approx 3.8R$, and lies above (x, y)-plane a height $d = z'_{\rm min} z_0 \sqrt{2}$, where $z'_{\rm min}$ expressed by equation (25) at incident point of the Bessel laser beam.The strength of the trap is gI_B. This simple pictorial model of the optical plasmon trap might be useful for further investigation and practical design in cold atom trap areas.

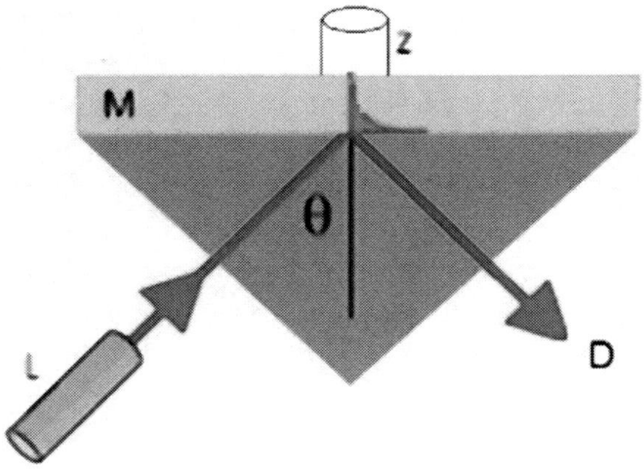

Figure 7. The pictorial model for the effective attractive optical plasmon potential introduced by evanescent effect. The trap is modeled by an elliptic cylinder. This ellipse expressed by equation (24) with average radius $R_D \approx 3.8R$ and lies above (x, y)-plane a height $d = z'_{\text{min}} z_0 \sqrt{2}$, where z'_{min} expressed by equation (25) at incident point of the Bessel laser beam. The strength of the trap is gI_B.

References

[1] J. A. Behr et al. 1997 *Phys. Rev. Lett.* **79** 357
[2] A. Fioretti et al. 1998 *Phys. Rev. Lett.* **80** 4402
[3] V. I. Balykin, K. Hakuta, Fam Le Kien, J. Q. Liang and M. Morinaga 2004 *Phys. Rev. A* **70** 011401
[4] F. L. Kien, V. I. V. I. Balykin, and K. Hakuta 2006 *Phys. Rev A* **74** 033412
[5] T. A. Chu, Do T. Nga, N. Tri Lan, N. A. Viet 2012 *IEEE Conference Series, IEEE Xplore*
[6] P. R. Giri 2008 *Int. J. Theor. Phys.* **47** 1776
[7] T. A. Chu, Do T. Nga, T. T. Thao, V. Thanh Ngo, and N. A. Viet 2011 *Mod. Phys. Lett. B* **25** 979
[8] Do T. Nga. N. T. Phuong Lan, N. A. Viet 2012 *Proc. 7th International Conference Photonics and Applications* 821-825
[9] Do T. Nga. N. T. Phuong Lan, and N.A. Viet 2014 *Journal of Physics: Conference Series* **537** 012014
[10] C. S. Yun et al. 2005 *J. Am. Chem. Soc.* **127** 3115
[11] C. J. Breshike, R. A. Riskowvski, G. F. Strouse 2013 *J. Phys. Chem. C* **117** 23942
[12] D. L. Andrews, D. S. Baradshaw 2004 *Eur. J. Phys.* **25** 845
[13] S. Bhowmick, S. Saini, V. B. Shenoy, B. Bagchi 2006 *J. Chem. Phys.* **125** 181102
[14] P. L. Stiles et al. 2008 *Annu. Rev. Anal. Chem.* **1** 601
[15] S. Al-Awfi 2012 *Sans Malaysiana* **41** 1461

Journal of Physics: Conference Series **627** (2015) 012018 doi:10.1088/1742-6596/627/1/012018

Second Quantization Model of Surface Plasmon Polariton at Metal Planar Surface

Dao Thi Thuy Nga[1], Nguyen Thi Phuong Lan[2] and Nguyen Ai Viet[3]

[1]Han Thuyen high school, Bac Ninh City, Bac Ninh Province, Vietnam

[2]Hanoi Pedagogical University No.2, Nguyen Van Linh Street, Phuc Yen Town, Vinh Phuc Province, Vietnam

[3]Institute of Physics. Vietnam Academy of Science and Technology, 10 Dao Tan, Thu Le, Ba Dinh, Ha Noi, Vietnam

E-mail: daothuyngaht@gmail.com

Abstract. We study on surface plasmon polariton modes at a metal planar surface. A model effective Hamiltonian with two interaction parameters is proposed in the second quantized representation for system of surface plasmons and photon. Using the Bogoliubov transformation technique, the dispersion relations of surface plasmon polariton was calculated. The surface plasmon photon vertexes also are found. A simple second quantized Hamiltonian of surface polaritons was obtained and could be useful for further investigations more complex systems.

1. Introduction

The plasmon is a quasiparticle resulting from the quantization of plasma oscillations. Plasmons are collective oscillations of the free electron gas density. Plasmons can couple with a photon to create another quasiparticle called a plasmon-polariton. Surface plasmons are those plasmons that are confined to surfaces and that interact strongly with light resulting in a polariton. They occur at the interface of a vacuum and material with a small positive imaginary and large negative real dielectric constant (usually a metal or doped dielectric). They play an important role in Surface Enhanced Raman Spectroscopy (SERS), Surface Plasmon Energy Transfer (SET) and Forster Resonance Energy Transfer (FRET), and have numerous applications in plasmonics, nanotechnology and modern medical treatments [1-3]. In this work we investigate quantum theory of surface plasmons and plasmon-polariton.

2. Drude model for bulk plasmon polariton

Consider metal as consisting of plasma (free electrons as a gas of free charge) in an ionic background (atoms). Existing the plasma will lead to waves in electron density which can be quantized by introducing the quasiparticle plasmon. Plasmons will again interact with surrounding electromagnetic field (photons) and found a new type of quasiparticle plasmon-polariton. The concept of plasmon-polariton is again useful as it was for exciton-polariton and phonon-polariton in condensed mater physics. In the frame work of phenomenological Drude model for metal, the metals dielectric constant is

$$\epsilon(\omega) = 1 - \frac{\omega_p^2}{\omega^2 - \omega_0^2 + i\gamma\omega}, \tag{1}$$

Journal of Physics: Conference Series **627** (2015) 012018 doi:10.1088/1742-6596/627/1/012018

where γ is the collision rate, ω_0 is the frequency of restoring force, ω_p is the bulk plasma frequency, $\omega_p = \sqrt{\frac{ne^2}{\epsilon_0 m}}$ in which n is the density, ϵ_0 is the vacuum permittivity and m is effective electron mass.
For simplicity we take the case of $\omega_0 = 0$ (no restoring force), and $\gamma \approx 0$ small collision rate, now the dielectric constant of a metal equals to

$$\epsilon(\omega) = 1 - \frac{\omega_p^2}{\omega^2}. \tag{2}$$

Bulk plasmon polariton is related to the bulk oscillations of plasma-like medium. The propagation of light through a bulk medium is governed by Maxwell's equations. In the case of plane wave with the wave vector $\vec{k}$, and a source-free isotropic medium, the equation div$\vec{D} = 0$ gives $\epsilon(\omega).(\vec{k}.\vec{E}) = 0$, which has the two solutions:
i) the first solution $\epsilon(\omega) = 0$ is associated to longitudinal modes $\omega = \omega_p$ where the gas of electrons oscillates as a collective whole;
ii) the second solution $(\vec{k}.\vec{E}) = 0$ is the transversality condition which yields, with the help of the Maxwell's equations rot$\vec{H} = \partial\vec{D}/\partial t$ and rot$\vec{E} = -\partial\vec{B}/\partial t$.

$$\epsilon(\omega)\omega^2 = c^2k^2, \tag{3}$$

where c is the speed of light in vacuum. Replacing equation (2) into equation (3) we found the dispersion relation for the transverse bulk plasmon polariton

$$\omega = \sqrt{\omega_p^2 + c^2k^2}. \tag{4}$$

Using plasmon energy ω_p as the energy unit and ω_p/c as the momentum unit, the dispersion relation of bulk plasmon polaritons in metals is presented in the figure 1.

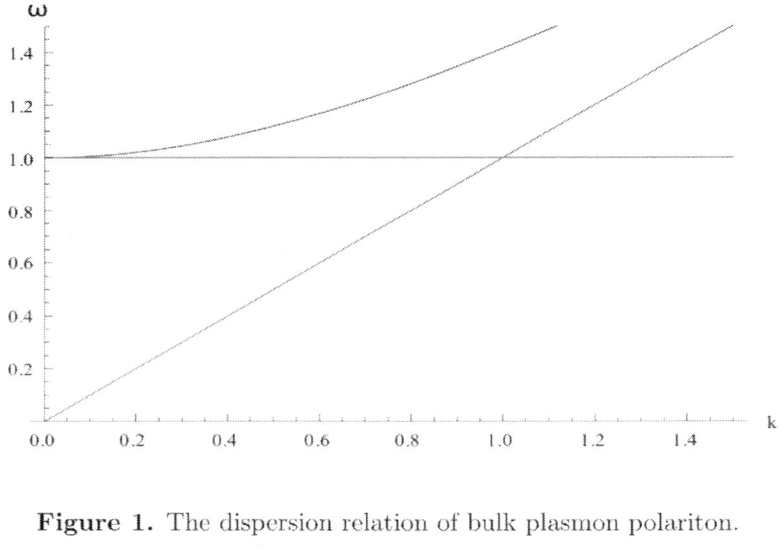

Figure 1. The dispersion relation of bulk plasmon polariton.

For $\omega > \omega_p$ bulk plasmon polaritons can propagate. In the large k limit, $\omega_p \to ck$ the photon in vacuum dispersion line. Frequency interval below plasma frequency $0 < \omega < \omega_p$ is the highly damped range, in which no propagation.

3. Drude model for surface plasmon polariton

Surface plasmon polariton (SPP) often referred simply as surface plasmons that are electromagnetic surface waves (photons) bound to the metal/dielectric interfaces. Physically, surface plasmon polaritons may be thought of as a collective oscillation of charge carriers in the conducting layer, with the result that the energy is transferred parallel to the interface.

The SPP is by definition the solution of Maxwell's equations describing an electromagnetic surface wave with lateral extensions evanescently decaying on either side of the interface. A plane wave with TE polarization (s-polarization) with its electric field parallel to the interface merely results in a motion of the electric charges parallel to the interface involving no restoring forces of the charges, and consequently no wave can propagate along the interface. For a TM-polarized wave (p-polarization) with its magnetic field vector parallel to the interface, the electric field has a non-null perpendicular component which results in the accumulation of electric charges at the interface. In this case, a restoring force is created because the charges are trapped inside the solid resulting in the formation of a propagating surface wave with the charges moving along the interface in the direction of propagation. The wave equation leads to a dispersion relation for plasmon polariton which depends on the dielectric constant. Solving the Maxwell's equations at the metal/dielectric planar interface (for simplicity we take the case of metal/vacuum interface), the dispersion relation for surface plasmons propagating on the interface when coupled to transverse magnetic (TM) illumination (photon) is

$$ck = \omega\sqrt{\frac{\epsilon(\omega)}{1+\epsilon(\omega)}}. \tag{5}$$

With Drude model, this dispersion relation can be rewritten in an explicit form

$$\omega_D(k) = \sqrt{c^2k^2 + \frac{\omega_p^2}{2} - \sqrt{c^4k^4 + \left(\frac{\omega_p^2}{2}\right)^2}}. \tag{6}$$

The dispersion relation of surface plasmon polariton $\omega_D(k)$ in the Drude model is presented in the figure 2.

For small k surface plasmon polariton is photon like, for large k it is surface plasmon like with strong coupling of mechanical and electromagnetic wave. In the large k limit, $\omega_p \to \omega_p/\sqrt{2}$ surface plasmon like dispersion relation. Frequency gap between the bulk and surface plasmon $\omega_p/\sqrt{2} < \omega < \omega_p$ is highly damped range, in which no propagation.

4. Second quantized Hamiltonian Model for surface plasmon polariton

In analogy the cases of exciton plariton and phonon polariton, we consider a Hamilatonian model for surface plasmon polariton in second quantization form

$$H = \sum_k H_k = \sum_k \omega_{\gamma k} a_k^+ a_k + E_{pk} c_k^+ c_k + g_k(a_k^+ c_k + c_k^+ a_k), \tag{7}$$

where a and a^+, c and c^+ are annihilation-creation photon and surface plasmon operators corresponding to momentum k, respectively; $\omega_{\gamma k}$=ck is the photon energy, $E_{pk} = \omega_p/\sqrt{2}$ is the surface plasmon energy.

We denote g_k is the plasmon-photon transition vertex (or coupling constant). This vertex is absence in traditional plasmon theory because plasmon is longitudinal excitation, while photon is transversal excitation. For surface plasmon, we consider plasmon-photon transition vertex is not equal zero, and being a main parameter of our theory. Using the Feynman diagram technique, we can express the plasmon polariton in the radom phase approximation (RPA) as the sum in the figure 3.

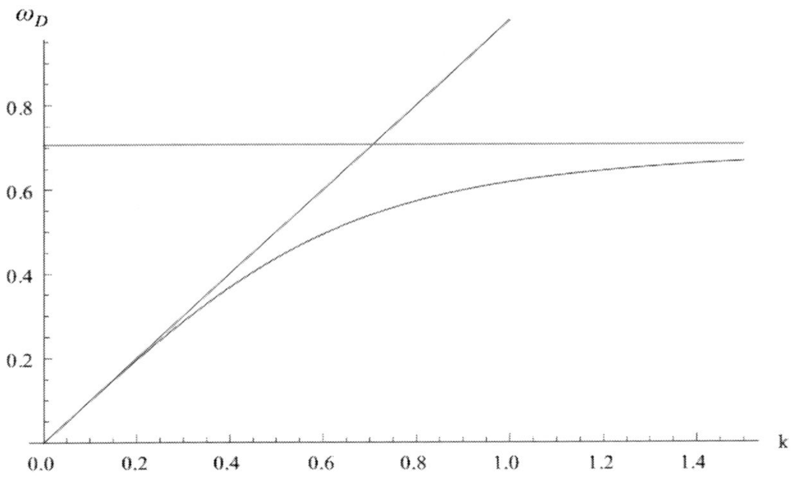

Figure 2. The dispersion relation of surface plasmon polariton $\omega_D(k)$ in the Drude model.

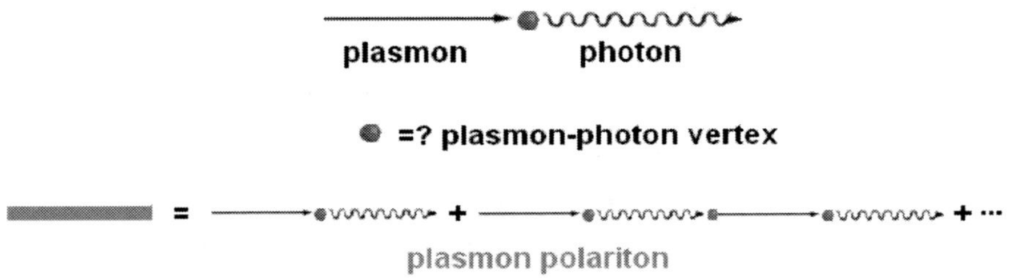

Figure 3. Feynman diagrams of vertex and propagation lines for plasmon polariton.

5. Bogoliubov Transformations

We use the Bogoliubov transformation technique taken from superconductivity theory to diagonalization plasmon polariton Hamiltonian

$$H_k = \sum_k \omega_{ik} \gamma_{ik}^+ \gamma_{ik}, \tag{8}$$

where γ_{ik} and γ_{ik}^+ are the annihilation and creation plasmon polariton operators with momentum k and i is branch number: $i = 1$ for lower and $i = 2$ for upper branch.
The transformations with condition $u_k^2 + \nu_k^2 = 1$ are

$$\gamma_{1k} = u_k a_k + \nu_k c_k, \; \gamma_{2k} = -\nu_k a_k + u_k c_k. \tag{9}$$

Using the commutation relations for annihilation and creation operators
$[a_k, a_k^+] = 1$, $[c_k, c_k^+] = 1$, $[\gamma_{ik}, \gamma_{ik}^+] = 1$, and zero for others, by standard calculation as in [4] we found the plasmon polariton energies

$$\omega_{ik} = \frac{1}{2}\left[(\omega_{\gamma k} + E_{pk}) + (-1)^i \sqrt{(\omega_{\gamma k} - E_{pk})^2 + 4g_k^2}\right]. \tag{10}$$

The upper branch is lying in the energy gap where damping is high. We take the lower branch for surface plasmon polariton dispersion relation

$$\omega_{SP}(k) = \omega_{1k} = \frac{1}{2}\left[\left(ck + \frac{\omega_p}{\sqrt{2}}\right) - \sqrt{\left(ck - \frac{\omega_p}{\sqrt{2}}\right)^2 + 4g_k^2}\right]. \tag{11}$$

The surface plasmon polariton dispersion relation ω_{SP} depends on wave vector k and coupling constant g is presented in the figure 4.

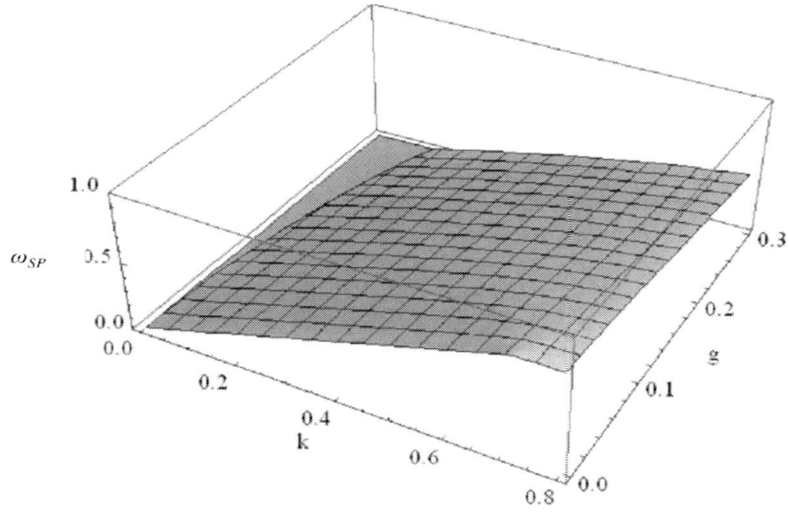

Figure 4. The surface plasmon polariton dispersion relation ω_{SP} depends on wave vector k and coupling constant g.

The surface plasmon polariton dispersion relation ω_{SP} depends on wave vector k and coupling constant $g = 0,\ 0.1,\ 0.2,\ 0.3$ is presented in the figure 5.
The full picture of bulk and surface plasmon polariton dispersion relation ω_{SP} depends on wave vector k and coupling constant $g = 0,\ 0.1,\ 0.2,\ 0.3$ with photon and surface plasmon guide lines is presented in the figure 6.
The figure 6 shows that the simple two bands model of surface plasmon polariton might be good in the most important bottom neck region but may be fare in the long wave limit $k = 0$.

6. Comparison with experiment

We compare our simple two-bands plasmon polariton model with reflection experiment in InSb [5] and Ag [6]. The comparison between the theoretical dispersion relation of surface plasmon

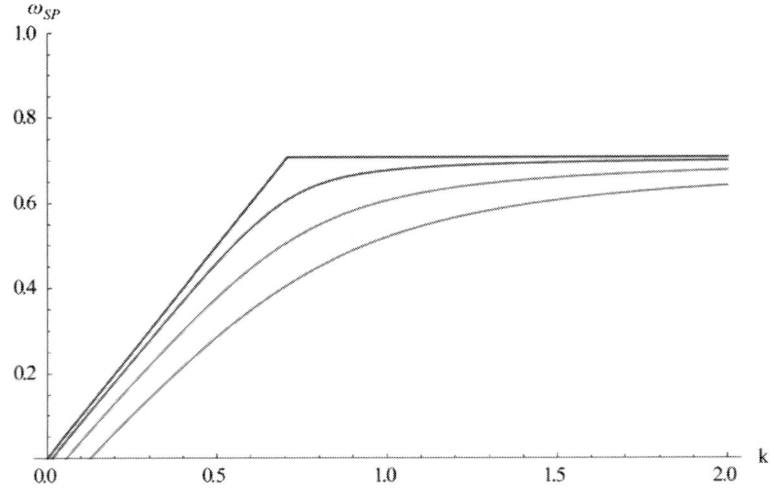

Figure 5. The surface plasmon polariton dispersion relation ω_{SP} depends on wave vector k and coupling constant $g = 0,\ 0.1,\ 0.2,\ 0.3$.

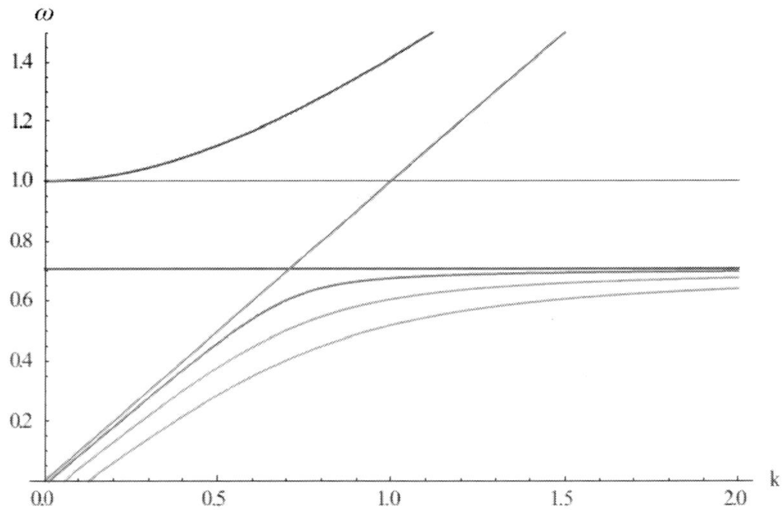

Figure 6. The full picture of bulk and surface plasmon polariton dispersion relation ω_{SP} depends on wave vector k and coupling constant $g = 0,\ 0.1,\ 0.2,\ 0.3$ with photon and surface plasmon guide lines.

polariton with the fitting value of plasmon-photon coupling constant $g_k = 0.1\,\omega_p$, and experiment data of a) InSb (dotted line), and b) Ag (dashed line, lower line is Drude and g with k-dependence) are presented in the figure 7.

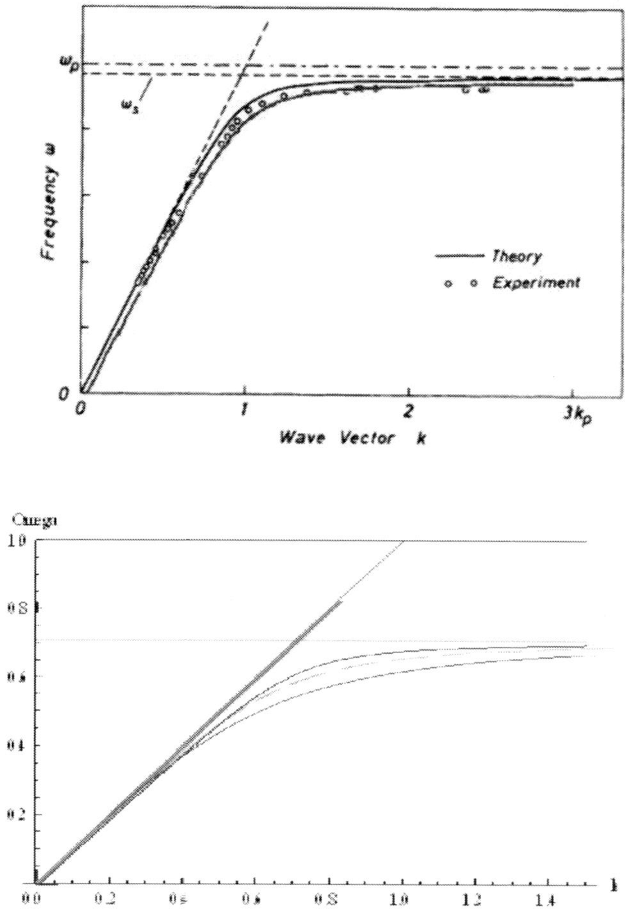

Figure 7. The comparison between the theoretical dispersion relation of surface plasmon polariton with the fitting value of plasmon-photon coupling constant $g_k = 0.1\,\omega_p$, and experiment data of a) InSb (dotted line), and b) Ag (dashed line, lower line is Drude and g with k-dependence).

We note here our theoretical obtained results from two bands quantum model with $g_k = 0.1\omega_p$ have quite good agreement with the experimental data in neck region but fare in the long wave limit. The Drude model and two bands quantum model with k-dependence of coupling constant give the same results and are good at low k limit but fare at bottom neck region.

7. Coupling constant with k-dependence

As mentioned above, the simple two bands model of surface plasmon polariton might be good in the most important neck region but may be fare in the long wave limit $k = 0$. In this part, we propose to overcome this problem by investigation the k-dependence of the plasmon-photon coupling constant.

Assuming the two dispersion relations of Drude $\omega_D(k)$ and our model $\omega_{SP}(k)$ are equal $\omega_D(k) = \omega_{SP}(k)$, we have an equation for finding the plasmon-photon coupling constant g_k

$$\sqrt{c^2k^2 + \frac{\omega_p^2}{2} - \sqrt{c^4k^4 + \left(\frac{\omega_p^2}{2}\right)^2}} = \frac{1}{2}\left[\left(ck + \frac{\omega_p}{\sqrt{2}}\right) - \sqrt{\left(ck - \frac{\omega_p}{\sqrt{2}}\right)^2 + 4g_k^2}\right]. \qquad (12)$$

The value of the plasmon-photon coupling constant g_k depends on wave vector k is plotted in the figure 8.

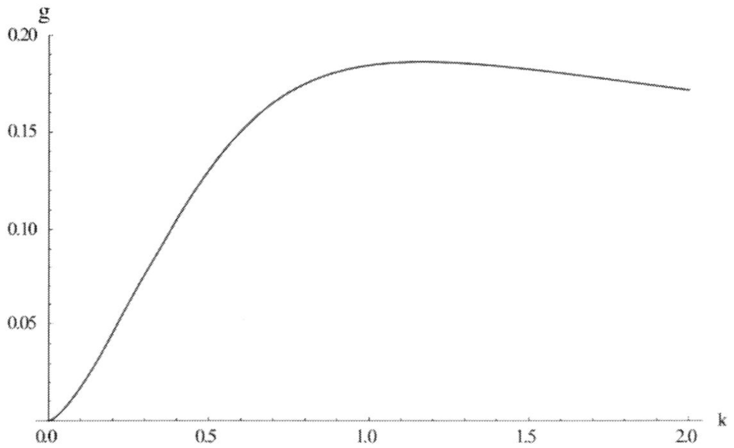

Figure 8. The value of the plasmon-photon coupling constant g_k depends on wave vector k.

At bottom neck point $g_k \approx 0.165\,\omega_p$. In the low k limit, $g_k \to 0$ when $k \to 0$, g_k reaches its approximate maximum value $0.185\,\omega_p$ in the bottom neck region, and decreases in the large k limit.

The common picture of bulk and surface plasmon polariton dispersion relation ω_{SP} depends on wave vector k with photon and surface plasmon guide lines is presented in the figure 9.

8. Conclusion and discussion

We proposed a simple two band model Hamiltonian of surface plasmon-polariton in second quantized representation. The main parameter of the model is surface plasmon-photon transition vertex (surface plasmon-photon coupling constant) g_k. We compare the lower surface plasmon polariton branch with experiment and have rather good agreement in the crossing region but not so well in the long wave limit. To overcome this problem we suggested a k-dependence of g_k and for that an explicit formula was found and investigated.

Our theoretical obtained results from two bands quantum model with $g_k = 0.1\,\omega_p$ have quite

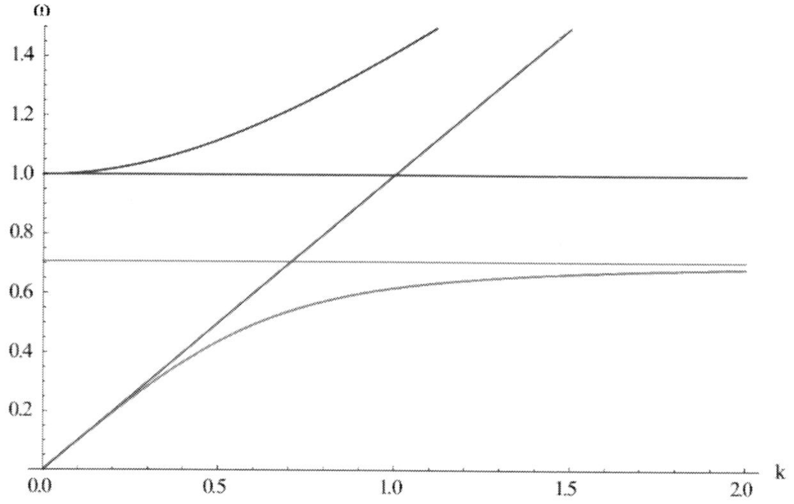

Figure 9. The common picture of bulk and surface plasmon polariton dispersion relation ω_{SP} depends on wave vector k with photon and surface plasmon guide lines.

good agreement with the experimental data in neck region but fare in the long wave limit. The Drude model and two bands quantum model with k-dependence of coupling constant g_k give the same results and are good at low k limit but fare at bottom neck region. Because surface plasmon polariton has a too large damping region from surface to bulk plasmon frequency in comparing with exciton and phonon polaritons, so the connection between the bulk and surface plasmon polariton dispersion relations still lacked. To investigate this problem we suggest constructing a new 3-band model for surface plasmon polariton, that will be investigated in the near future.

References

[1] M. A. Garcia 2011 *Journal of Physics D: Applied Physics* **44** 28
[2] Homola, Jiri 2006 *Surface Plasmon Resonance Based Sensors* (Berlin: Springer)
[3] Wei Xie, Penghe Qiu, and Chuanbin Mao 2011 *J. Mater. Chem.* **21** 5190
[4] N. B. An, N. V. Hieu, N. T. Thang, T. L. Thang, and N. A. Viet 1981 *Ann. Phys.* **131** 149
[5] B. Fischer, N. Marschall, and H.J. Queisser 1973 *Surf. Sci.* **34** 50
[6] S. Hellstrom 2012 *Exciton-plasmon interaction in semiconductor-metal nanostructures* (PhD Thesis, Stockholm, Sweden)

Journal of Physics: Conference Series **627** (2015) 012019 doi:10.1088/1742-6596/627/1/012019

Influence of confined acoustic phonons on the Radioelectric field in a Quantum well

Do Tuan Long and Nguyen Quang Bau

Faculty of Physics, Hanoi University of Science, Vietnam National University
334 Nguyen Trai, Thanh Xuan, Hanoi, Vietnam

E-mail: dotuanlong@hus.edu.vn

Abstract. The influence of confined acoustic phonons on the Radioelectric field in a quantum well has been studied in the presence of a linearly polarized electromagnetic wave and a laser radiation. By using the quantum kinetic equation for electrons with confined electrons – confined acoustic phonons interaction, the analytical expression for the Radioelectric field is obtained. The formula of the Radioelectric field contains the quantum number m characterizing the phonons confinement and comes back to the case of unconfined phonons when m reaches to zero. The dependence of the Radioelectric field on the frequency of the laser radiation, in case of confined acoustic phonons, is also achieved by numerical method for a specific quantum well AlGaAs/GaAs/AlGaAs. Results show that the Radioelectric field has a peak and reaches saturation as the frequency of the laser radiation increases.

1. Introduction

Recently, the study of the low-dimensional semiconductor systems (quantum wells, doping superlattices, compositional superlattices, quantum wires, quantum dots,...) has been increasingly interested. In these systems, the motion of electrons is restricted [7], and thus, leading to unusual kinetic properties of semiconductors such as: the absorption coefficient of electromagnetic waves [2], the Acoustoelectric effect [5], the Hall effect [6], the Radioelectric effect [8],...

The Radioelectric effect, the effect of drag of charge carriers by electromagnetic waves, has been studied in [1, 4, 8]. However, these studies only considered the electrons - unconfined phonons interaction while recent works show the remarkable contribution of phonons confinement in these kinetic properties [2, 3]. So, in this work, we study the influence of confined acoustic phonons on the Radioelectric field in a quantum well.

In literature, there are some methods to solve this problem such as: the Boltzmann equation, the Kubo-Mori method, the quantum kinetic equation method. The last approach, because of its simplicity and accuracy, is chosen to use in this work.

The article is organized as follows: we briefly describe the Radioelectric effect, the Hamiltonian of electrons in quantum well and present basic formulae for the calculation. Numerical results and discussion are given in section 3. And the final section shows remarks and conclusions.

Journal of Physics: Conference Series **627** (2015) 012019 doi:10.1088/1742-6596/627/1/012019

2. The Radioelectric field in a quantum well under the influence of confined acoustic phonons

We consider a quantum well with high infinite potential subjected to a linearly polarized electromagnetic field: $\vec{E}(t) = \vec{E}(e^{-i\omega t} + e^{i\omega t}), \vec{H}(t) = \frac{1}{c}[\vec{n}, \vec{E}(t)]$ ($\hbar\omega \ll \bar{\varepsilon}$, $\bar{\varepsilon}$ is an average carrier energy) and a laser radiation $\vec{F}(t) = \vec{F}\sin\Omega t$ ($\Omega\tau \gg 1$, τ is the characteristic relaxation time). The Hamiltonian of the confined electrons - confined acoustic phonons system can be expressed as:

$$\mathrm{H}=\sum_{n,\vec{p}_\perp} \varepsilon_n(\vec{p}_\perp - \frac{e}{\hbar c}\vec{A}(t))a^+_{n,\vec{p}_\perp} c_{n,\vec{p}_\perp} + \sum_{m,\vec{q}_\perp} \hbar\omega_{m,\vec{q}_\perp} b^+_{m,\vec{q}_\perp} b_{m,\vec{q}_\perp} + $$

$$+ \sum_{n,n',m,\vec{p}_\perp,\vec{q}_\perp} C_{m,\vec{q}_\perp} I^m_{nn'} a^+_{n',\vec{p}_\perp+\vec{q}_\perp} a_{n,\vec{p}_\perp}(b_{m,\vec{q}_\perp} + b^+_{m,-\vec{q}_\perp}) \tag{1}$$

where $a^+_{n,\vec{p}_\perp}$ and $a_{n,\vec{p}_\perp}$ ($b^+_{m,\vec{q}_\perp}$ and $b_{m,\vec{q}_\perp}$) are the creation and annihilation operators of electron (phonon) respectively; $\varepsilon_n(\vec{p}_\perp) = \frac{\hbar^2}{2m^*}(\vec{p}_\perp^{\,2} + \frac{\pi^2 n^2}{L^2})$ is the electron energy; m^* is the effective mass of an electron; $\hbar\omega_{m,\vec{q}_\perp}$ is the confined acoustic phonon energy; $\vec{A}(t) = \frac{c}{\Omega}\vec{F}\cos(\Omega t)$ is the vector potential of the laser radiation.

$|C_{m,\vec{q}_\perp}|^2 = \frac{\hbar^2\xi^2}{2\rho v_s}\sqrt{q_\perp^2 + q_m^2}$ is the confined electron - confined acoustic phonon interaction constant, here ξ, ρ, v_s are the deformation potential constant, the mass density and the sound velocity, respectively.

$I^m_{nn'} = \frac{2}{L}\int\limits_0^L(\eta(m)\cos\frac{m\pi z}{L} + \eta(m+1)\sin\frac{m\pi z}{L})\sin\frac{n\pi z}{L}\sin\frac{n'\pi z}{L}dz$ is the electron form factor, $\eta(m) = 1$ if m is even and $\eta(m) = 0$ if m is odd, m is the quantum number characterizing the phonons confinement.

Using the Heisenberg equation of motion, we establish the quantum kinetic equation for electrons distribution function $f_{n,\vec{p}_\perp}(t) = \left\langle a^+_{n,\vec{p}_\perp} a_{n,\vec{p}_\perp}\right\rangle_t$:

$$\frac{\partial f_{n,\vec{p}_\perp}(t)}{\partial t} + (e\vec{E}(t) + e\vec{E}_0 + \hbar\omega_c[\vec{p}_\perp, \vec{h}], \frac{\partial f_{n,\vec{p}_\perp}(t)}{\hbar\vec{p}_\perp}) =$$

$$= \sum_{n,n',m,\vec{p}_\perp,\vec{q}_\perp} \frac{2\pi}{\hbar^2}|C_{m,\vec{q}_\perp}|^2|I^m_{nn'}|^2 N_{m,\vec{q}_\perp} \sum_{l=-\infty}^{+\infty} J_l^2(\frac{e\vec{F}}{m^*\Omega^2}, \vec{q}_\perp)[f_{n',\vec{p}_\perp+\vec{q}_\perp}(t) - f_{n,\vec{p}_\perp}(t)]\times$$

$$\times[\delta(\varepsilon_{n',\vec{p}_\perp+\vec{q}_\perp} - \varepsilon_{n,\vec{p}_\perp} + \hbar\omega_{m,\vec{q}_\perp} - l\hbar\Omega) + \delta(\varepsilon_{n',\vec{p}_\perp+\vec{q}_\perp} - \varepsilon_{n,\vec{p}_\perp} - \hbar\omega_{m,\vec{q}_\perp} - l\hbar\Omega)] \tag{2}$$

here $\vec{h}(t) = \frac{\vec{H}(t)}{H}, \omega_c = \frac{eH}{m^*}$ is the cyclotron frequency.

We look for the electrons distribution function as:

$$f_{n,\vec{p}_\perp}(t) = f_0(\varepsilon_{n,\vec{p}_\perp}) - \vec{p}_\perp\vec{\chi}_{n,\vec{p}_\perp}(t)\frac{\partial f_0(\varepsilon_{n,\vec{p}_\perp})}{\partial\varepsilon_{n,\vec{p}_\perp}} \tag{3}$$

where $\vec{\chi}_{n,\vec{p}_\perp}(t) = \vec{\chi}_0(\varepsilon_{n,\vec{p}_\perp}) + \vec{\chi}(\varepsilon_{n,\vec{p}_\perp})e^{-i\omega t} + \vec{\chi}^*(\varepsilon_{n,\vec{p}_\perp})e^{i\omega t}$ and $f_0(\varepsilon_{n,\vec{p}_\perp})$ is the equilibrium electron distribution function.

We substitute (3) into (2) and obtain:

$$\vec{\chi}_0(\varepsilon_{n,\vec{p}_\perp}) = \frac{e\hbar}{m}\tau(\varepsilon_{n,\vec{p}_\perp})\vec{E}_0$$

$$\vec{\chi}(\varepsilon_{n,\vec{p}_\perp}) = \frac{e\hbar}{m}\frac{\tau(\varepsilon_{n,\vec{p}_\perp})}{1 - i\omega\tau(\varepsilon_{n,\vec{p}_\perp})}\vec{E}$$

The total current density is presented in the form:

$$\vec{j}_{tot} = \int\limits_0^\infty \vec{R}(\varepsilon)d\varepsilon \tag{4}$$

Journal of Physics: Conference Series **627** (2015) 012019 doi:10.1088/1742-6596/627/1/012019

where

$$\vec{R}(\varepsilon) = \sum_{n,\vec{p}_\perp} \frac{e\hbar}{m} \vec{p}_\perp f_{n,\vec{p}_\perp}(t)\, \delta\left(\varepsilon - \varepsilon_{n,\vec{p}_\perp}\right) \tag{5}$$

Multiplying both sides of the equation (2) by $\frac{e\hbar}{m}\vec{p}_\perp \delta\left(\varepsilon - \varepsilon_{n,\vec{p}_\perp}\right)$ and sum over $n, \vec{p}_\perp$. Then, putting the root $\vec{R}(\varepsilon)$ of the resulting equation into (4) and assuming the sample in all direction is opened, regarding the electron gas is completely degenerate $f_0\left(\varepsilon_{n,\vec{p}_\perp}\right) = \theta\left(\varepsilon_F - \varepsilon_{n,\vec{p}_\perp}\right)$. We obtain the expression of the Radioelectric field after some manipulation:

$$E_{0y} = \frac{2\omega_c \tau(\varepsilon_F)}{1+\omega^2\tau^2(\varepsilon_F)} \frac{a+c}{a+b} \left(E_x h_z - E_z h_x\right) \tag{6}$$

where: $\varepsilon_n = \frac{\hbar^2\pi^2}{2m^* L^2} n^2$; $\varepsilon_{n'} = \frac{\hbar^2\pi^2}{2m^* L^2} n'^2$

$$a = \sum_n \frac{e^2}{\pi\hbar^2}\left(\varepsilon_F - \varepsilon_n\right) \tag{7}$$

$$\text{b} = \sum_{n,n',m} \frac{e^4\xi^2 kTF^2}{4\pi\hbar^7\Omega^4\rho v_s^2} |I_{nn'}^m|^2 \left\{ \left(\varepsilon_F - \varepsilon_n\right)\left(4\varepsilon_F - 3\varepsilon_{n'} - \varepsilon_n - \hbar\Omega\right)\tau(\varepsilon_F) - \right.$$

$$\left. - (\varepsilon_F - \varepsilon_n')(4\varepsilon_F - \varepsilon_n + \hbar\Omega)\tau(\varepsilon_F + \hbar\Omega) \right\} \tag{8}$$

$$\text{c} = \sum_{n,n',m} \frac{e^4\xi^2 kTF^2}{4\pi\hbar^7\Omega^4\rho v_s^2} |I_{nn'}^m|^2 \left\{ \left(\varepsilon_F - \varepsilon_n\right)\left(4\varepsilon_F - 3\varepsilon_{n'} - \varepsilon_n - \hbar\Omega\right)\tau(\varepsilon_F) \frac{1-\omega^2\tau^2(\varepsilon_F)}{1+\omega^2\tau^2(\varepsilon_F)} - \right.$$

$$\left. - \left(\varepsilon_F - \varepsilon_{n'}\right)\left(4\varepsilon_F - \varepsilon_n + \hbar\Omega\right) \frac{\tau^2\left(\varepsilon_F + \hbar\Omega\right)}{\tau\left(\varepsilon_F\right)} \frac{1 - \omega^2\tau\left(\varepsilon_F\right)\tau\left(\varepsilon_F + \hbar\Omega\right)}{1 + \omega^2\tau^2\left(\varepsilon_F + \hbar\Omega\right)} \right\} \tag{9}$$

Formula (6) shows the dependence of the Radioelectric field on parameters of the system: the frequency Ω and amplitude F of the laser radiation, the frequency ω of the linearly polarized electromagnetic wave, the temperature T and and especially the quantum number m characterizing the phonons confinement. When m goes to zero, we obtain the results in the case of unconfined phonons.

3. Numerical results and discussions

In order to clarify the influence of confined acoustic phonons, we estimate numerical values of the Radioelectric field and graph in cases of confined phonons and unconfined phonons. The parameters used in this calculation are as follows: $m^* = 0.067m_0$ (m_0 is the mass of a free electron), $\xi = 13.5\text{eV}$, $\rho = 5.32\text{gcm}^{-1}$, $v_s = 5378\text{ms}^{-1}$, $\varepsilon_F = 30\text{meV}$, $\tau(\varepsilon_F) = 10^{-12}\text{s}$, $T = 35K$, $L = 30\text{nm}$.

Figure 1 shows the dependence of the Radioelectric field on the frequency of the laser radiation in cases of confined and unconfined phonons. It can be seen that nonlinear dependences of the Radioelectric field on the laser radiation frequency are similarly in both cases, the Radioelectric field has a peak and reaches saturation when the laser frequency increases. However, the maximum value of the Radioelectric field is much greater than that in the case of unconfined phonons. This behaviour of the Radioelectric field could be explain by the effect of confined acoustic phonons, the appear of the quantum number m in the electron form factor $I_{nn'}^m$ changes remarkably the intensity of the Radioelectric field. For more details, we numerically calculate the Radioelectric field for different values of the quantum number m characterizing the phonons confinement. In figure 2, we can see that when m increases, the Radioelectric field is also increasing.

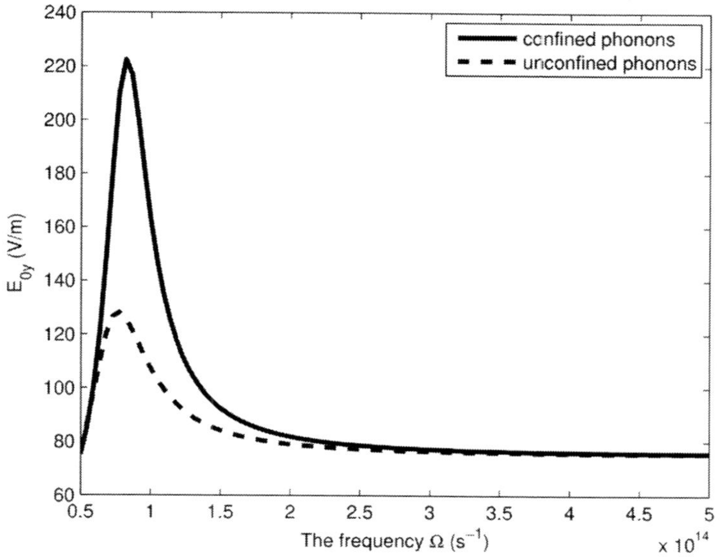

Figure 1. *The dependence of the Radioelectric field on the frequency of the laser radiation in the case of confined phonons (solid line) and the case of unconfined phonons (dashed line). Here* $E = 5 \times 10^4 V/m$ *and* $\omega = 4.5 \times 10^{11} s^{-1}$.

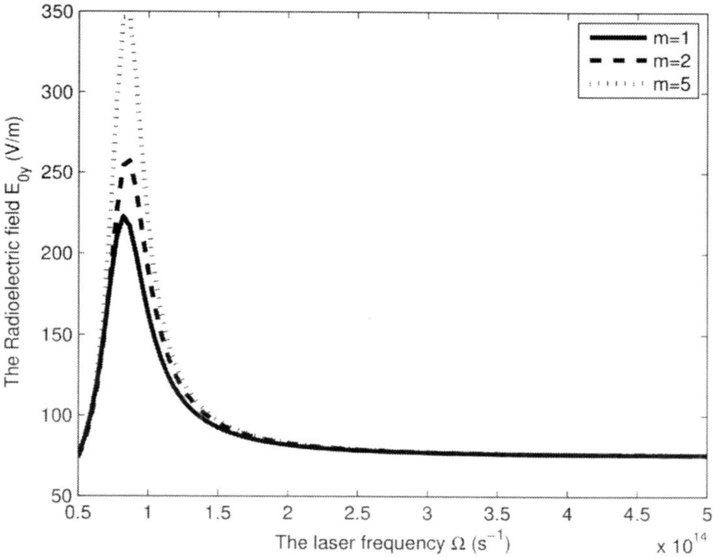

Figure 2. *The dependence of the Radioelectric field on the frequency of the laser radiation for different values of the quantum number m characterizing the phonons confinement:* $m = 1$*(solid line),* $m = 2$*(dashed line) and* $m = 5$*(dotted line). Here* $E = 5 \times 10^4 V/m$ *and* $\omega = 4.5 \times 10^{11} s^{-1}$.

4. Conclusions

In this paper, the influence of confined acoustic phonons on the Radioelectric field in a quantum well has theoretically studied. We have computed the analytical expression for the Radioelectric field, numerically calculated and graphed the theoretical result for the AlGaAs/GaAs/AlGaAs quantum well, and compared with the case of unconfined phonons. Results show that the formula of the Radioelectric field contains the quantum number m characterizing the phonons confinement and comes back to the case of unconfined phonons when m goes to zero. The Radioelectric field has a peak and reaches saturation when the laser frequency increases in both cases, but the Radioelectric field is much greater when phonons are confined. Hence, the phonons confinement is the causes of surprising changes to the Radioelectric field in the quantum well.

Acknowledgment

This work is completed with financial support from the QGTD 12.01 and Vietnam International Education Development - Project 911.

References

[1] Shmelev G M, Tsurkan G I and Epshtein E M 1982 *Phys. Stat. sol.* (b) **109**.
[2] Bau N Q, Hung L T and Nam N D 2010 *Journal of Electromagnetic Waves and Applications* **24** 1751–61.
[3] Bau N Q, Hung D M and Hung L T 2010 *Progress In Electromagnetics Research Letters* **15** 175–185.
[4] Bui Duc Hung, Nguyen Vu Nhan, Luong Van Tung, Nguyen Quang Bau 2012 *Proc. Natl. Conf. Theor. Phys.* **37** 168–173.
[5] Nguyen Quang Bau, Nguyen Van Hieu and Nguyen Vu Nhan 2012 *Superlattices and Microstructure* **52** 921–930.
[6] Nguyen Quang Bau and Bui Dinh Hoi 2012 *Journal of the Korean Physical Society* **60** 59–64.
[7] Vasilopoulos P, Charbonneau M and Van Vliet C M 1987 *Phys. Rev.* B **35** 1334–44.
[8] Kryuchkov S V, Kukhar E I and Sivashova E S 2008 *Physics of the Solid State* **50** 1150–56.

Quantum coherent π-electron rotations in a non-planar chiral molecule induced by using a linearly polarized UV laser pulse

Hirobumi Mineo[1] and Yuichi Fujimura[2]

[1]Institute of atomic and molecular science, No.1 Section 4, Roosevelt road, Taipei, 115, Taiwan

[2]Department of Applied Chemistry, Institute of Molecular Science and Center for Interdisciplinary Molecular Science, National Chiao Tung University, Hsinchu 30010, Taiwan

E-mail: mineo@gate.sinica.edu.tw

Abstract. We propose an ultrafast quantum switching method of π-electron rotations, which are switched among four rotational patterns in a nonplanar chiral aromatic molecule (*P*)-2,2'-biphenol and perform the sequential switching among four rotational patterns which are performed by the overlapped pump-dump laser pulses. Coherent π-electron dynamics are generated by applying the linearly polarized UV pulse laser to create a pair of coherent quasi-degenerated excited states. We also plot the time-dependent π-electron ring current, and discussed ring current transfer between two aromatic rings.

1. Introduction

Recently organic electronics and devices have been actively studied [1-8] because of its potential usefulness. Particularly, π-electron dynamics in aromatic ring molecules are focused very much since π-electron shows high reactivity. The quantum simulations of π-electron ring current were performed in a higher symmetric planar molecules, Mg-porphyrin [9, 10]. Here the degenerated electronic states are excited by using the circularly polarized UV laser pulses. The rotational direction of the resultant "incoherent" π-electron ring current depends on whether the right or left circularly polarized laser pulse is applied, and the current is unidirectional. On the other hand the linearly polarized UV laser pulses can create a "coherent" π-electron ring current in a chiral aromatic molecule with lower symmetry. Kanno et. al. carried out numerical simulations of the π-electron rotation for 2,5-dichloro[n](3,6) pyrazinophane [11, 12], which is a planar chiral aromatic molecule. The rotational direction of π-electron ring current is determined by whether the superposition of quasi-degenerate coherent states excited by linearly polarized UV laser pulse is in-phase or out-phase. The results of numerical simulation indicates that the chiral aromatic molecule can be a candidate of the ultrafast switching device.

We have also studied coherent π-electron dynamics of a nonplanar chiral aromatic molecule (*P*)-2,2'-biphenol [13, 14]. This is a nonplanar chiral aromatic molecule with axial chirality, which has two aromatic rings (called *L* and *R* hereafter) linked through a single chemical bridge bond. We have classified four possible rotational patterns of π-electron rotations [13, 14] as CC, AC, CA, and AA where C (A) represents clockwise (anticlockwise) rotational direction. Because of the nonplanar structure of (*P*)-2,2'-biphenol, the resultant π-electron rotation and angular momentum are two dimensional. This indicates that nonplanar chiral aromatic molecules like (*P*)-2,2'-biphenol may serve as a two-dimensional ultrafast switching device. For more effective quantum control of the ultrafast sequential switching among the above four rotational patters, the overlapped pump and dump pulses with a properly selected relative phase and a laser polarization direction are necessary in addition to the time delay between pump and dump pulses.

The paper is organized as follows. In section 2 we mention an optimized geometry and electronic excited states of (*P*)-2,2'-biphenol, and in section 3 the four directional patterns of the π-electron rotations and angular momentum on each ring *L* and *R* are shown. In section 4 we plot time-dependent behaviors of π-electron ring currents on both aromatic rings and a bridge current between two rings, and demonstrate an ultrafast two-dimensional quantum switching of π-electron angular momentum in addition. In section 5 we summarize and conclude this paper.

2. Optimized geometry and electronic excited states of (*P*)-2,2'-biphenol

Here we consider a (*P*)-2,2'-biphenol, which is a simple, and a real nonplanar chiral aromatic molecule with two covalently bended aromatic rings. The ground state geometry of (*P*)-2,2'-biphenol was optimized by using the DFT B3LYP level of theory in the GAUSSIAN09 program. The energy levels of the electronic excited states were calculated at the optimized ground state geometry by the TDDFT B3LYP level of theory. The 6-31G+(d,p) basis sets were used in our calculations. For generation of coherent angular momentum and ring current of (*P*)-2,2'-biphenol, we focus on the three optically allowed excited states (a, b_1 and b_2) The electronic excited energies at the a, b_1 and b_2 states, which were calculated at the optimized ground state geometry using the TD-DFT B3LYP level of theory are 6.67, 6.78, and 6.84 eV, respectively. The transition dipole moments are shown in Fig. 1. The transition dipole moment from the ground state to excited state a (A), $\vec{\mu}_{ga}(=(0,0,-0.77))$ [a.u.], is parallel to the Z-axis, and the others, $\vec{\mu}_{gb1}(=(0.08,1.93,0))$ and $\vec{\mu}_{gb2}(=(1.24,-0.34,0))$ [a.u.], are nearly orthogonal to each other in the *XY*-plane as shown in Fig. 1.

Coherent two electronic excited states $(a \pm b_1)$, $(a \pm b_2)$ and $(b_1 \pm b_2)$ are generated by applying a UV laser pulse with a properly selected polarization direction. Here $(\alpha+\beta)$ ($(\alpha-\beta)$) represents a phase between two electronic excited states is in-phase (out-phase). For example, for $(a \pm b_2)$ electronic coherence, the condition for a linearly polarized UV laser pulse with polarization vector $\vec{e}^{(+)}_{ab2}$ is given as $\vec{\mu}_{ga} \bullet \vec{e}^{(\pm)}_{ab2} = \pm\vec{\mu}_{gb2} \bullet \vec{e}^{(\pm)}_{ab2}$, and $\vec{\mu}_{gb1} \bullet \vec{e}^{(\pm)}_{ab2} = 0$.

Thus, even laser pulse overlaps with an a, b_1 and b_2, $(a+b_2)$ or $(a-b_2)$ electronic coherent state can be selectively generated.

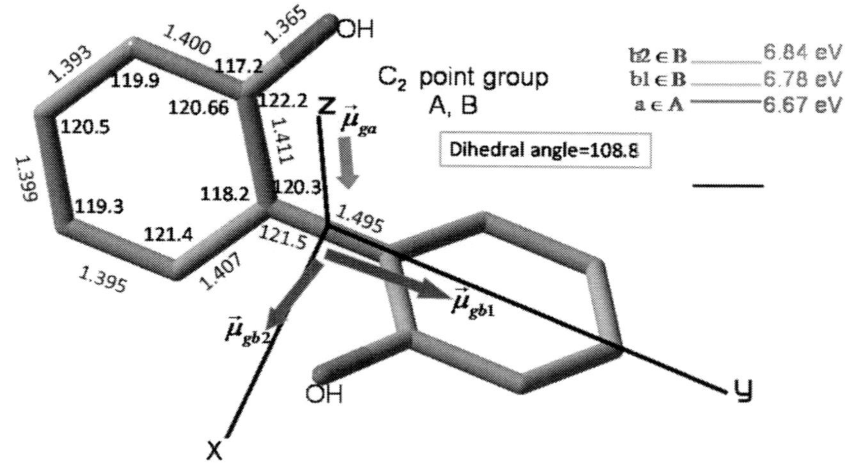

Figure 1. Geometrical structure of (*P*)-2,2'-biphenol with transition dipole moments between the ground state and three electronic excited states *a, b*1 and *b*2. The unit of bond lengths is in angstrom ($=10^{-10}$ m).

3. Methods

The initial directional patterns of π-electron rotations on each ring *L* and *R* are schematically depicted in Fig. 2. Here in the upper figure the selective electronic excited states with in- or out- phase are initially prepared by the UV laser pulse with the polarization direction $\vec{e}^{(\pm)}_{ab1}$, $\vec{e}^{(\pm)}_{ab2}$ and $\vec{e}^{(\pm)}_{b1b2}$. From Fig. 2 in the uper it is observed that in the cases (i) and (ii) in which a superposition of same irreducible representations: $(b_1 \pm b_2)$, π-electron rotations on each ring *L* and *R* oscillate in the same direction, whereas the bridge bond current is permanently zero. Whereas in other cases ring currents on each ring *L* and *R* oscillate in the opposite direction, as a characteristic feature a bond current between two rings *L* and *R* is generated.

Lower figure of Fig. 2 shows the resultant angular momentum $\vec{l} = \vec{l}_L + \vec{l}_R$ corresponding to each initial directional pattern of ring currents respectively. The π-electron angular momentum $\vec{l}_L$ ($\vec{l}_R$) is perpendicular to the *L* (*R*) phenol ring, and in the cases (i) and (ii), i.e., in which a superposition of $(b_1 \pm b_2)$, the resultant total angular momentum (denoted as l_Z) is parallel to the Z axis, whereas in other cases in which a superposition of different irreducible representations: $(a \pm b_1)$ or $(a \pm b_2)$, the resultant total angular momentum (denoted as l_X) is parallel to the X axis, in the XZ plane.

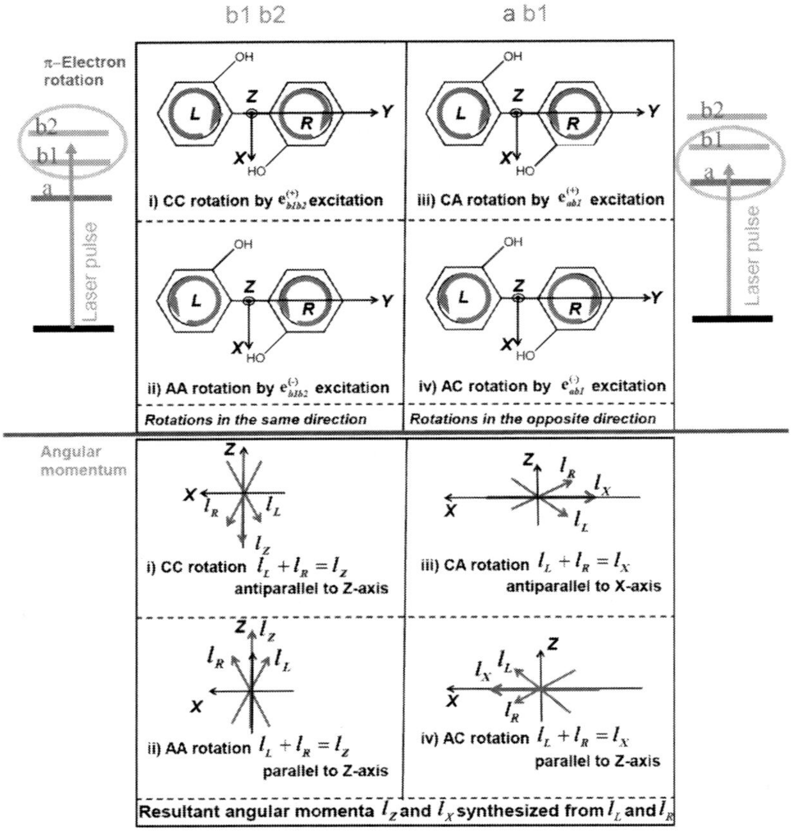

Figure 2. The initial directional patterns of π-electron rotations on each ring L and R induced by selective coherent electronic excited states $(a \pm b_1)$, $(a \pm b_2)$ and $(b_1 \pm b_2)$ are shown in the upper figure. For example, CA rotation means directions of π-electron rotations on L and R are clockwise and anticlockwise respectively. (b) Resultant angular momentum l_X and l_Z corresponding to each pattern of π-electron rotations are plotted respectively. l_L (l_R) denotes the angular momentum created on the L (R) ring.

4. Results and discussions

Here we first show the results of the π-electron ring current generated on the (P)-2,2' biphenol, and demonstrate two-dimenional ultrafast quantum switching of the angular momentum.

In Fig. 3(a) directions of the ring current J_L (J_R) on L (R) phenol ring, and the bridge bond current are defined. In figure 3(b) we shine the π-pulse gaussian-shaped LP laser pulse with the amplitude 1.2 GV/m and pulse duration 64 fs. Here polarization direction of the laser pulse is chosen as in-phase direction to coherently excite to quasi-degenerate electronic excited states b_1 and b_2. It is observed that ring currents J_L and J_R oscillate in the same phase, whereas the bridge bond current J_B is permanently zero. On the other hand, in figure 3(c) we shine the π-pulse gaussian-shaped LP laser pulse with the amplitude 2.5 GV/m and pulse duration 38 fs. Similarly to Fig. 3(b), the polarization direction is in-phase direction to excite to quasi-degenerate electronic excited states a and b_1. J_L and J_R oscillate in the opposite phase, and as a characteristic feature J_B is generated, and the amplitude of transferred current is smaller than the amplitudes of ring currents. Transfer of ring currents will occur if the created coherent excited stated states belong to different irreducible representations A and B.

A three-dimensional plot of the resultant angular momentum switching based on the sequential four-step scheme is seen in Fig. 4. It can be seen from Fig. 4a that the angular momenta were successfully controlled by the pulses depicted in Fig. 4b, i.e., both the rotational axis (parallel to the Z or X axis) and the rotational direction around the axis (clockwise or anticlockwise) were satisfactorily controlled by the sequential four-step process. In Fig. 4b, the quantum control at each switching step was carried out by using overlapped pump and dump pulses with specific polarization directions.

For the present effective ultrafast quantum switching, overlap between the pump and dump pulses is essential: the resultant electric field is rotated as an elliptically polarized one in the overlapped region, and the electric field forces the rotating π electrons to induce the reverse rotation that occurs in this overlapped region. As a result, the electronic angular momentum or coherence between two electronic quasi-degenerate states is erased.

(a)

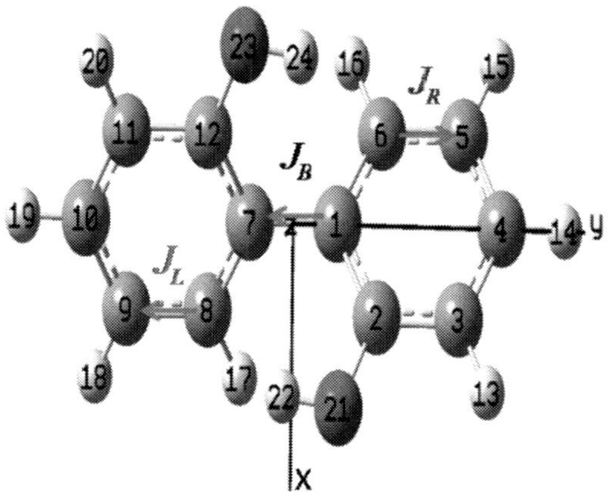

(b)

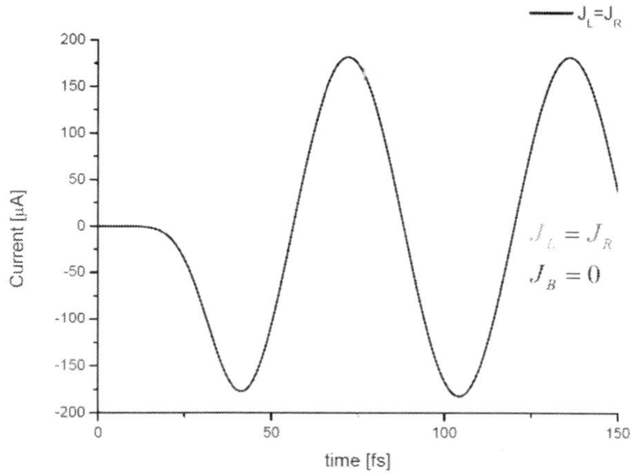

(c)

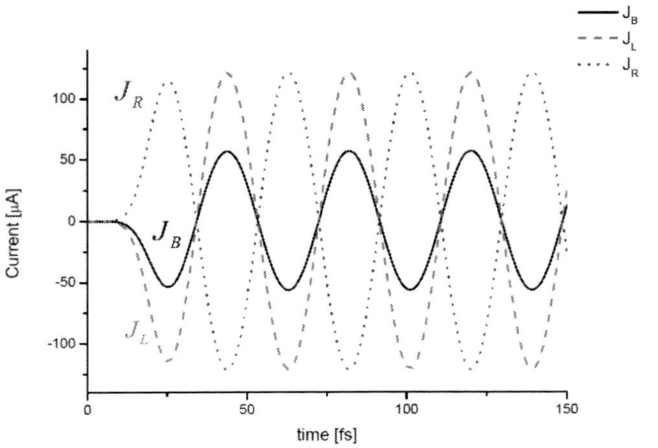

Fig. 3. (a) Definitions of directions of the ring current J_L (J_R) on L (R) phenol ring, and the bridge bond current J_B are given. (b) Time evolution of ring currents J_L, J_R and bridge bond current J_B are plotted. Here coherent electronic excited states b_1 and b_2 are created with in-phase relative coherence by a LP laser pulse with the amplitude 1.2 GV/m and pulse duration 64 fs. (c) Time evolution of ring currents J_L, J_R and bridge bond current J_B are plotted. Here coherent electronic excited states a and b_1 are created with in-phase relative coherence by a LP laser pulse with the amplitude 2.5 GV/m and pulse duration 38 fs.

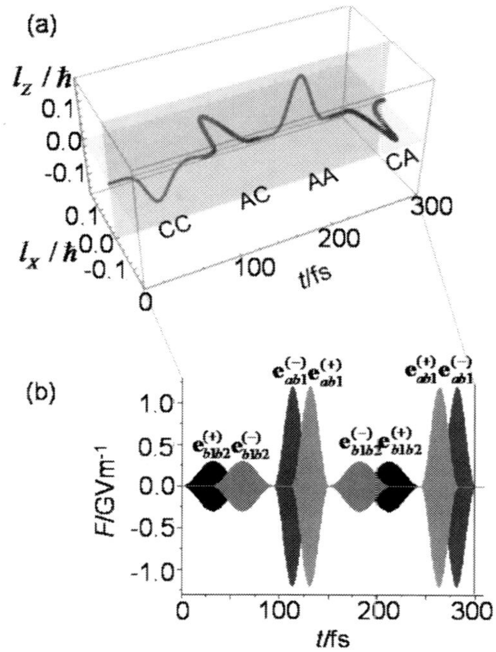

Fig. 4. (a) Sequential four-step switching of π-electron rotations in (*P*)-2,2′-biphenol. (b) Electric fields of the sequence of overlapped pump and dump pulses.

5. Summary and conclusions

In conclusion we have calculated a coherent π electron angular momentum and ring current in chiral molecule, (*P*)-2,2'-biphenol by using the UV linearly polarized laser pulse. We showed time-dependent coherent ring currents and angular momentum for three types of in-phase electronic coherent excited states $(a+b_1)$, $(a+b_2)$ with B symmetry and (b_1+b_2) with A symmetry in the point group C_2. As a characteristic feature for electronic coherent states $(a+b_1)$ and $(a+b_2)$, a non-zero bridge bond current appears and oscillates between two phenol rings, while for electronic coherence (b_1+b_2) the bridge bond current disappears permanently. The direction of resultant angular momentum for electronic coherences $(a+b_1)$ and $(a+b_2)$ is parallel to Z-axis, while for electronic coherence (b_1+b_2) is parallel to X-axis. The initial directions of ring currents and angular momenta are determined from the initial phases of electronic coherences.

As an example of ultrafast quantum switching we have demonstrated an ultrafast two-dimensional switching of π-electron angular momentum in (*P*)-2,2'-biphenol using a sequence of overlapped pump-dump pulses with different phase and polarization direction. The key points for the multi-dimensional quantum switching is to coherently excite quasi-degenerate electronic states of a nonplanar chiral aromatic ring molecules with covalently bonded aromatic rings. Finally it would be fascinating to show methods of directly measuring the coherent electronic dynamics of a nonplanar chiral aromatic molecule.

Acknowledgement

Journal of Physics: Conference Series **627** (2015) 012020 doi:10.1088/1742-6596/627/1/012020

This work was supported by the National Science Council of Taiwan (Grant No. 102-2112-M-001-003-MY3). H.M. would like to thank Professor J.-L. Kuo for his critical comments and support.

References

[1] F. Remacle and R. D. Levine, Phys. Rev. A **83**, 013411 (2011).
[2] I. S. Ulusoy and M. Nest, J. Am. Chem. Soc. **133**, 20230 (2011).
[3] Y. Fujimura and H. Sakai, *Electronic and Nuclear Dynamics in Molecular Systems* (World Scientific, Singapore, 2011), p. 117.
[4] F. Krausz and M. Ivanov, Rev. Mod. Phys. 81, 163 (2009).
[5] M. Nest, F. Remacle, and R. D. Levine, New J. Phys. 10, 025019 (2008).
[6] F. Remacle, M. Nest, and R. D. Levine, Phys. Rev. Lett. 99, 183902 (2007).
[7] P. Krause, T. Klamroth, and P. Saalfrank, J. Chem. Phys. 123, 074105 (2005).
[8] P. Kral, T. Seideman, J. Chem. Phys. 123, 184702 (2005).
[9] I. Barth and J. Manz, Angew. Chem., Int. Ed. **45**, 2962 (2006).
[10] I. Barth, J. Manz, Y. Shigeta, and K. Yagi, J. Am. Chem. Soc. **128**, 7043 (2006).
[11] M. Kanno, H. Kono, and Y. Fujimura, Angew. Chem., Int. Ed. **45**, 7995 (2006).
[12] M. Kanno, K. Hoki, H. Kono, and Y. Fujimura, J. Chem. Phys. **127**, 204314 (2007).
[13] H. Mineo, Y. Teranishi, M. Yamaki, M. Hayashi, S.H. Lin, and Y. Fujimura, J. Am. Chem. Soc. **134**, 14279 (2012).
[14] H. Mineo, S.H. Lin, and Y. Fujimura, J. Chem. Phys. **138**, 074304 (2013).

IWTCP-2 and NCTP-39 IOP Publishing
Journal of Physics: Conference Series **627** (2015) 012021 doi:10.1088/1742-6596/627/1/012021

First, Second Quantization and Q-Deformed Harmonic Oscillator

Man Van Ngu[1], **Ngo Gia Vinh**[2], **Nguyen Tri Lan**[3], **Luu Thi Kim Thanh**[4] **and Nguyen Ai Viet**[3]

[1]Hung Yen Industrial Colleage, Hung Yen Province, Vietnam
[2]Bac Ninh Department of Education and Training, Bac Ninh Province, Vietnam
[3]Institute of Physics, Vietnam Academy of Science and Technology, 10 Dao Tan street, Ba Dinh, Hanoi, Vietnam
[4]Hanoi Pedagogical University No.2, Vinh Phuc Province, Vietnam

E-mail: cnttcs2@gmail.com

Abstract. Relations between the first, the second quantized representations and deform algebra are investigated. In the case of harmonic oscillator, the axiom of first quantization (the commutation relation between coordinate and momentum operators) and the axiom of second quantization (the commutation relation between creation and annihilation operators) are equivalent. We shown that in the case of q-deformed harmonic oscillator, a violence of the axiom of second quantization leads to a violence of the axiom of first quantization, and inverse. Using the coordinate representation, we study fine structures of the vacuum state wave function depend in the deformation parameter q. A comparison with fine structures of Cooper pair of superconductivity in the coordinate representation is also performed.

1. Introduction

In the last decade, studies of quantum algebras (or groups) have a great attention in modern physics and mathematics. They motivated by the passage from classical physical systems to quantum systems. Despite concept of algebras was suggested long before the discovery of quantum mechanics in the beginning of the last century, and an important achievement was mad by Heisenberg, the Heisenberg deformed algebra still rare applications in comparison with wave representation (Schrodinger wave equation, or the first quantization), and particle representation (occupation numbers approach, or second quantization) of quantum mechanics. Recently, quantum group and deformed Heisenberg algebras with q-deformed harmonic oscillator have been a subject of intensive investigation. This approach is found some applications in various branches of physics and chemistry [1–7]. The method of q-deformed quantum mechanics was developed on the base of Heisenberg commutation relation (the Heisenberg algebra). The main parameter of this method is the deformation parameter q usually considered to variety in the range $0 < q < 1$, and the models have been constructed that the behavior of studying objects reduce to theirs standard counterparts as $q \to 1$. In this work we investigate relations between the first, the second quantized representations of quantum mechanics and deform algebra. In the case of harmonic oscillator, the axiom of first quantization (the commutation relation between coordinate and momentum operators) and the axiom of second quantization (the commutation relation between creation and annihilation operators) are equivalent, so harmonic oscillator is a good object for investigation the link between the two representations of quantum mechanics. We shown that in the case of q-deformed harmonic oscillator, a violence of the axiom of second quantization leads to a violence of the axiom of first quantization, and inverse. Using the coordinate representation, we study fine structures of the vacuum state wave function depend on the deformation parameter q. A comparison with fine structures of Cooper pair of superconductivity in the coordinate representation also is performed. In this work we use the atomic unit system with $\hbar = c = k_B = 1$.

2. Relation between the first and the second quantized representations of quantum mechanics

One of most important behaviors of nature is the wave-particle dualism. In the first quantization, classical particles turned to some waves (or fields in general). Example electrons can be expressed by wave function $\Psi_e(\mathbf{r},t)$, satisfied the wave equation Schrodinger. The regular habitants in first quantization are plane waves. Vacuum of the first quantization (the home land of wave habitants) is four-dimension flat Hilbert space. which is denoted by the ket-bra $|0\rangle = (\mathbf{r}, t)$. The axioms of the first quantization are the commutation relations between coordinate and momentum $[\hat{x}, \hat{p}] = -i$. In the second quantization, waves (or fields) become particle (quasi-particles in general). Example light wave become photon. The regular habitants in second quantization are particles (quasi-particles). Vacuum of the second quantization (the home land of particle habitants) is the Fock space (or occupation number space), which is denoted by the round-bras $|0)$. The axioms of the second quantization are the commutation relations between annihilation and creation operators (commutation for bosons and anti-Commutation for fermions). Relations between the first and the second quantized representations of quantum mechanics are expressed in the table 1.

Name	First quantization	Second quantization	Connection
Vacuum	Hilbert space $\|0\rangle$	Fock space $\|0\rangle$	
Habitant	Wave	Particle	Fourier transform
Main variable	Coordinate x Momentum p	Level number N	
Axiom	Commutation relations of coordinate and momentum $[\hat{x}, \hat{p}] = i$	Relations for annihilation and creation operators Commutation (boson)$[a, a^+] = 1$ Anti-Commutation (fermions) $\{b, b^+\} = 1$	
Evolution variable	Time t	Temperature T	Conjugate
Physical constant	Planck constant $\hbar$	Boltzmann constant k_B	
Main uncertainty or chaos	Uncertainty principle $\triangle x \triangle p = \hbar$, $\triangle t \triangle E = \hbar$	Zero-level (vacuum) fluctuation Zero temperature fluctuation	
Main state	Steady	Equilibrium	
Behavior	Spreading	Rotation (spin)	
Distribution& Statistics	Uniform (free) Wave packet (bound)	Fermi-Dirac (fermion) Bose-Einstein (boson)	
System main principle	Minimal energy	Maximal entropy	
Main presentation	Schrodinger equation	Hamiltonian operator	Harmonic oscillator

Table 1. Relations between the first and the second quantized representations of quantum mechanics.

Note here harmonic oscillator is a good bridge between the first and the second quantized representations of quantum mechanics.

3. Harmonic oscillator ($q = 1$)

For simplicity we put the mass of oscillator is $m = 1$, and take the case of one dimensional harmonic oscillator. In the first quantization representation the Hamiltonian of harmonic oscillator is

$$H_0 = \frac{\omega}{2}\left(\hat{p}^2 + \hat{x}^2\right). \tag{1}$$

where ω is the oscillation frequency, $\hat{x}$ and $\hat{p} = -i\frac{\mathrm{d}}{\mathrm{d}x}$ are the coordinate and momentum operators, which satisfied the commutative relation $[\hat{x}, \hat{p}] = \hat{x}\hat{p} - \hat{p}\hat{x} = i$. In the second quantization representation the Hamiltonian operator of harmonic oscillator is

$$H_0 = \frac{\omega}{2}\left(a_0 a_0^+ + a_0^+ a_0\right), \tag{2}$$

where a_0^+, a_0 are creation and annihilation operators, which satisfied the commutation relation $\left[a_0, a_0^+\right] = a_0 a_0^+ - a_0^+ a_0 = 1$. Energy spectrum of harmonic oscillator has the form

$$E_{0n} = \frac{\omega}{2}(2n + 1), \tag{3}$$

where $n = 0, 1, 2, 3, \ldots$ are the integers. The creation and annihilation operators of harmonic oscillator can be expressed in terms of coordinate and momentum operators as

$$a_0 = \frac{(\hat{x} + i\hat{p_x})}{\sqrt{2}}, \tag{4}$$

$$a_0^+ = \frac{(\hat{x} - i\hat{p_x})}{\sqrt{2}}. \tag{5}$$

From the definition of the annihilation operator $a_0|0)$, the ground state wave function of harmonic oscillator Ψ_0 satisfies the equation

$$a_0|0) = 0 = \left[\frac{(\hat{x} + i\hat{p_x})}{\sqrt{2}}\right]\Psi_0 = \frac{1}{\sqrt{2}}(\hat{x} + \frac{\mathrm{d}}{\mathrm{d}x})\Psi_0. \tag{6}$$

The solution of this equation gives us the ground state wave function of harmonic oscillator Ψ_0 in a Gaussian form

$$\Psi_0(x) = Ce^{-\frac{x^2}{2}}, \tag{7}$$

where C is the normalization constant. The ground state wave function of harmonic oscillator $C\Psi_0$ with is plotted in the figure 1.

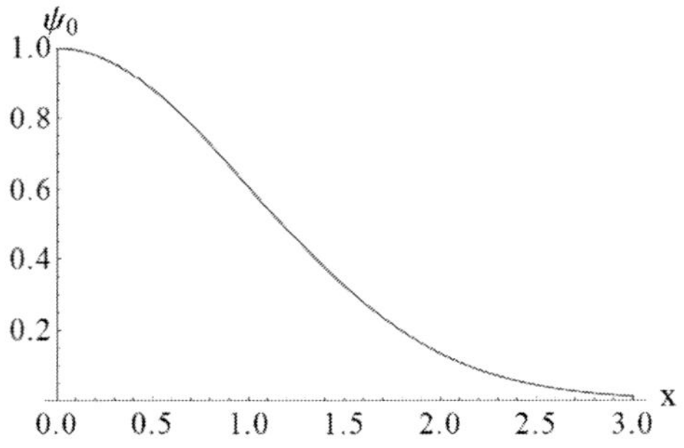

Figure 1. The ground state wave function of harmonic oscillator Ψ_0 has a Gaussian form.

Easily to check that

$$\left(0|a_0^+ a_0|0\right) = 0, \tag{8}$$

and

$$\langle 0|a_0^+ a_0|0\rangle = \langle 0|\Psi_0^+\Psi_0|0\rangle = \langle 0|C^2 e^{-x^2}|0\rangle = C^2 \int_{-\infty}^{\infty} dx e^{-x^2} = C^2\Pi^{\frac{1}{2}} \neq 0, \tag{9}$$

therefore, but $\langle 0|a^+a|0\rangle \neq 0$ so $|0) \neq |0\rangle$. We note here the vacuums of the first and the second quantized representations of quantum mechanics are not the same. Introducing density operator $\rho_0 = \Psi_0^+\Psi_0$ and consider that the density fluctuation of vacuum in first quantization (Hilbert space) is corresponding the zero level oscillation $\left(n = \frac{1}{2}\right)$ of harmonic oscillator in the second quantization Fock space (see Casimir effect)

$$\langle 0|\rho_0|0\rangle = \frac{1}{2}, \tag{10}$$

from that we can calculate the value of constant C

$$C = \left(2^{\frac{1}{2}}\Pi^{\frac{1}{4}}\right)^{-1}. \tag{11}$$

4. q-deformed harmonic oscillator

Creation and annihilation operators of q-deformed harmonic oscillator satisfied the commutation relation

$$[a, a^+]_q = aa^+ - qa^+a = 1, \tag{12}$$

where q is deformation parameter taking values in $[0, 1]$. Introducing the new deformation parameter α taking values in $[0, \infty]$

$$\alpha = \sqrt{-\frac{\ln q}{2}}. \tag{13}$$

In the second quantized representation, the Hamiltonian operator of q-deformed harmonic oscillator is

$$H = \frac{\omega}{2}\left(aa^+ + a^+a\right). \tag{14}$$

Energy spectrum of q-deformed harmonic oscillator has the form

$$E_n = \frac{\omega}{2}\left([n] + [n+1]\right), \tag{15}$$

where $[n] = \frac{1-q^n}{1-q}$ are the d-integer, they differ from natural numbers. The values of ratio $\frac{[n]}{n}$ depend on deformation parameter q are presented in the figure 2.

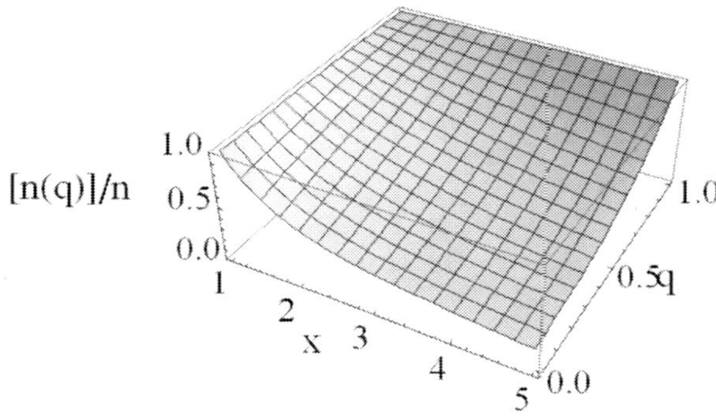

Figure 2. The values of ratio $\frac{[n]}{n}$ depend on deformation parameter q.

This ratio tends to zero $\frac{[n]}{n} \to 0$ when $q \to 0$. The maximum value of $[n]_{\max} \to \infty$ when $q \to 1$ and we return to case of standard harmonic oscillator. The creation and annihilation operators of q-deformed harmonic oscillator can be expressed in terms of coordinate and momentum operators as

$$a = \frac{\exp\left(-2i\alpha x\right) - \exp\left(i\alpha \frac{\mathrm{d}}{\mathrm{d}x}\right)\exp\left(-i\alpha x\right)}{\sqrt[-i]{1 - \exp\left(-2\alpha^2\right)}}, \tag{16}$$

$$a^+ = \frac{\exp\left(2i\alpha x\right) - \exp\left(i\alpha x\right)\exp\left(i\alpha \frac{\mathrm{d}}{\mathrm{d}x}\right)}{\sqrt[i]{1 - exp\left(-2\alpha^2\right)}}. \tag{17}$$

Denote $\varepsilon = 1 - q$, the energy spectrum of q-deformed harmonic oscillator becomes quadratic

$$E_n = \omega\left(n + \frac{1}{2} - \frac{n^2}{2}\varepsilon + O\left(\varepsilon^2\right)\right). \tag{18}$$

As above, the solution of the equation $a|0\rangle = 0$

$$a|0\rangle = \frac{\exp\left(-2i\alpha x\right) - \exp\left(i\alpha \frac{\mathrm{d}}{\mathrm{d}x}\right)\exp\left(-i\alpha x\right)}{\sqrt[i]{1-\exp\left(-2\alpha^2\right)}}, \tag{19}$$

gives us the ground state wave function Ψ_0 of q-deformed harmonic oscillator in the form

$$\Psi_0\left(x\right) = C\exp\left(-\frac{x^2}{2}+\frac{3}{2}i\alpha x\right), \tag{20}$$

In the case of $q \to 1$, $\alpha \to 0$ we return to case of standard harmonic oscillator. Introduce density operator $\rho = \Psi_0^+\Psi_0$ of q-deformed harmonic oscillator, because $\Psi_0^+\Psi_0 = \psi_0^+\psi_0$ therefore $\rho = \rho_0$, so that $\langle 0|\rho|0\rangle = \langle 0|\rho_0|0\rangle = \frac{1}{2}$. Values of density vacuum fluctuations are the same for both harmonic oscillator and q-deformed harmonic oscillator. Real part ReΨ_0 and imaginary part Im Ψ_0 of ground state wave function Ψ_0 are presented in figure 3.

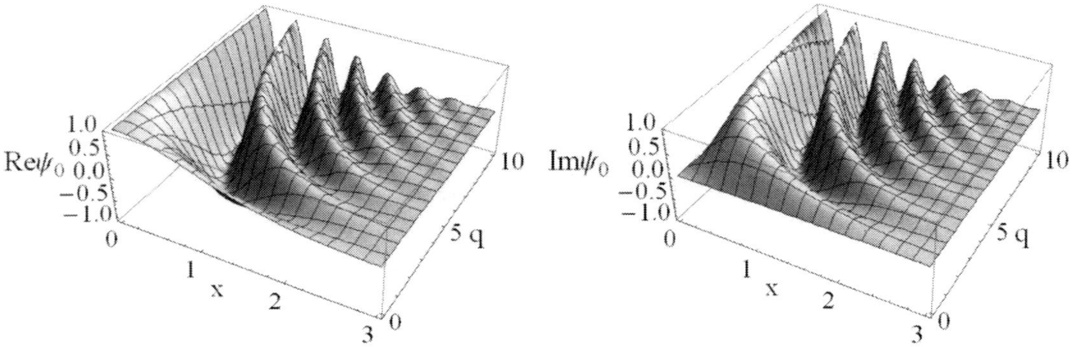

Figure 3. Real part ReΨ_0 and imaginary part ImΨ_0 of ground state wave function Ψ_0 are presented in figure 3.

At $q = 1$, back to harmonic oscillator case with ReΨ_0 has a Gaussian form, and Im$\Psi_0 = 0$.

5. Comparison with Cooper pairs in superconductivity

From condensed matter physics we have learn that the real part of wave function plays important role. In this part we try to explore the meaning of real part ReΨ_0 of ground state wave function Ψ_0 of q-deformed harmonic oscillator. For comparison we take the well known case of Cooper pairs in superconductivity. In a long history of the BCS theory, the Cooper pair usually analyzed in the momentum-space. The first investigation Cooper pair in coordinate-space was done in the work [8], where was shown that this leads to a spherically symmetrical quasi-atomic wave function, with an identical onion-like layered structure for each of the electrons constituting the Cooper pair. The internal structure of Cooper pair wave function Ψ_C (also called the singlet pair function or the Gorkov function) is given by

$$\Psi(r) \propto \int \frac{cos\left(k_F r + e'r'\right)d\varepsilon'}{\sqrt{\varepsilon'^2+1}} \approx cos\left(k_F r\right)\int \frac{cos\left(\varepsilon' r'\right)d\varepsilon'}{\sqrt{\varepsilon'^2+1}} = cos\left(k_F r\right)K_0\left(\frac{r}{\pi\xi_0}\right), \tag{21}$$

Journal of Physics: Conference Series **627** (2015) 012021 doi:10.1088/1742-6596/627/1/012021

where k_F is the Fermi wave vector at the top of the Fermi sea, K_0 is the zero-order modified Bessel function with an asymptotic form that is similar to an exponential for large x. The wave function Ψ_C of Cooper pair is plotted in the figure 4.

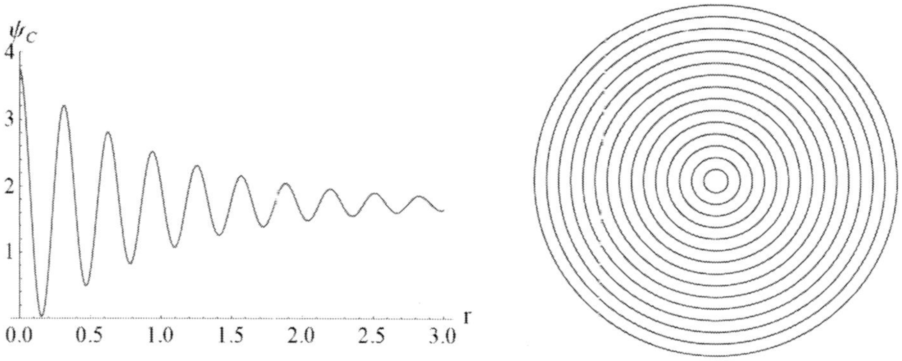

Figure 4. The wave function Ψ_C of Cooper pairs in superconductivity.

We can realize that the real part $\mathrm{Re}\Psi_0$ of ground state wave function Ψ_0 of q-deformed harmonic oscillator (see figure 5) is very similar the wave function Ψ_C of Cooper pair.

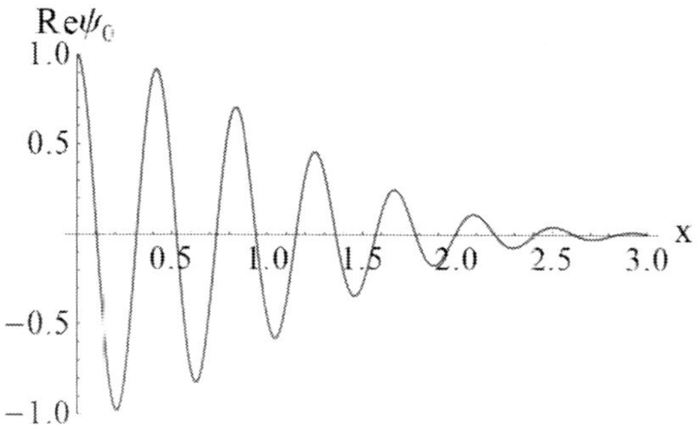

Figure 5. The real part $\mathrm{Re}\Psi_0$ of ground state wave function Ψ_0 of q-deformed harmonic oscillator $\alpha = 10$.

The similarity between $\mathrm{Re}\Psi_0$ of q-deformed harmonic oscillator and the wave function Ψ_C of Cooper pair might be not accidental and will be a subject for further investigation.

6. Uncertainty relation for q-deformed harmonic oscillator

The coordinate-momentum uncertainty relation is an important character of quantum world. This uncertainty for q-deformed harmonic oscillator is

$$\triangle x \triangle p = \frac{E}{\omega} = \frac{([n] + [n+1])}{2}.$$

For the case of harmonic oscillator $(\triangle x \triangle p)_0 = \frac{(2n+1)}{2}$. The ratio of $K = \frac{\triangle x \triangle p}{(\triangle x \triangle p)_0}$ of uncertainty for q-deformed harmonic oscillator and uncertainty for harmonic oscillator is plotted in the figure 6.

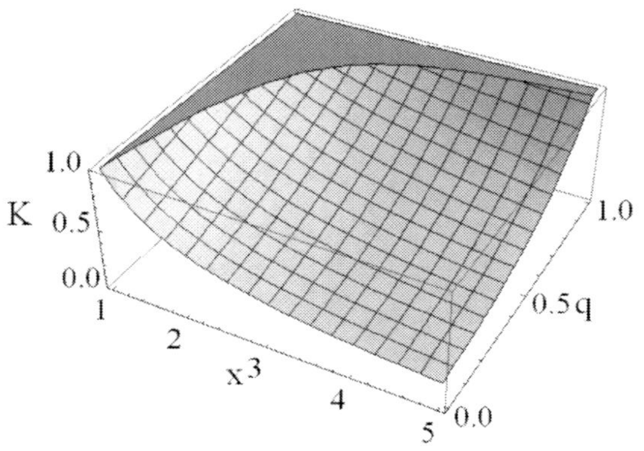

Figure 6. The ratio of $K = \frac{\triangle x \triangle p}{(\triangle x \triangle p)_0}$ of uncertainty for q-deformed harmonic oscillator and uncertainty for harmonic oscillator.

We note here uncertainty for q-deformed harmonic oscillator is less than the analogous value for standard harmonic oscillator for all n, except $n = 0$. At $n = 0$, the coordinate-momentum uncertainty is minimum and given by $1/2$.

7. Discussion

The main results of this work are the table1 showing the relations between the first and the second quantized representations of quantum mechanics with an attention that harmonic oscillator is a good bridge between them, also the investigation deformation parameter q depending on the most important physical characters of q-deformed harmonic oscillator in explicit coordinate representations, such as annihilation and creation operators, q-deformed integer, the ground wave function, coordinate-momentum uncertainty relation. Introducing density operators, we explored the physical meaning of wave function of ground state of the two types oscillators with and absence deformation, and shown that the vacuums of first and second quantization are not the same but the values of vacuum density fluctuation or zero level oscillation are equal for both. We shown that the real part $\mathrm{Re}\Psi_0$ of ground state wave function Ψ_0 of q-deformed harmonic oscillator is very similar the wave function Ψ_C of Cooper pair in superconductivity. This similarity might be not accidental and will be a subject for further investigation. We noted that uncertainty for q-deformed harmonic oscillator is less than the analogous value for standard harmonic oscillator for all n, except $n = 0$, which in contrast to standard quantum mechanics that uncertainty of harmonic oscillator is minimal value. The physical meaning of this also will be a subject for further investigation.

References

[1] V. V. Eremin, A. A. Meldianov 2008 *arXiv:0810.1967v1.*
[2] S. Abe, C. Beck, E. G. D. Cohen 2007 *Phys. Rev.* E **76** 031102.
[3] S. Abe 2009 *Phys. Rev.* E **79** 041116.
[4] A. Lavagno, P. P. Swamy 2010 *Physica.* A **389** 933.

Journal of Physics: Conference Series **627** (2015) 012021 doi:10.1088/1742-6596/627/1/012021

[5] A. Algin, M. Senay 2012 *Phys. Rev.* E **85** 041123.
[6] F. M. Andrade, E. O. Silva 2013 *Phys. Lett.* B **719** 467 .
[7] V. M. Tkachuck 2013 *Phys. Rev.* A **86** 062112 .
[8] A. M Kadin 2007 *JSNM* **20** 285.

Journal of Physics: Conference Series **627** (2015) 012022 doi:10.1088/1742-6596/627/1/012022

Q-Deformed Harmonic Oscillator and Morse-like Anharmonic Potential

Ngo Gia Vinh[1], Man Van Ngu[2], Nguyen Tri Lan[3], Luu Thi Kim Thanh[4] and Nguyen Ai Viet[3]

[1]Bac Ninh Department of Education and Training, Bac Ninh Province, Vietnam
[2]Hung Yen Industrial College, Hung Yen Province, Vietnam
[3]Institute of Physics, Vietnam Academy of Science and Technology, 10 Dao Tan Str., Ba Dinh, Hanoi, Vietnam
[4]Hanoi Pedagogical University No.2, Vinh Phuc Province, Vietnam

E-mail: vinhphongvan@gmail.com

Abstract. Connection between q-deformed harmonic oscillator and Morse-like anharmonic potential is investigated. It is well known that the potential of harmonic oscillator is parabolic. We have shown that the potential of q-deformed harmonic oscillator could be a Morse-like anharmonic potential. The relation between the deformation parameter q and the set of parameters of Morse-like anharmonic potential was found. We also investigated the partition function and some thermodynamic properties of q-deformed harmonic oscillator.

1. Introduction

Recently, quantum group and deformed Heisenberg algebras with q-deformed harmonic oscillator have been a subject of intensive investigation. This approach has found a number of applications in various branches of physics and chemistry [1–7]. In context, the method of q-deformed quantum mechanics was developed on the base of Heisenberg commutation relation i.e. the Heisenberg algebra. The main parameter in this method is the deformation parameter q varying in the range $0 < q < 1$, and the models have been constructed so that the physical behaviours of studying objects reduce to theirs standard counterparts as $q \to 1$.

In other hand, the well–known Morse and Morse–like potentials have an important role in describing the interaction of atoms in diatomic and even polyatomic molecules [8–12] in atomic and molecular physics. Despite its quite simple form, the Morse potential describes very well the vibrations of diatomic molecules. This is because that four-particle complex system (two heavy atomic nuclei with positive charge and two light electrons with negative charge) could be reduced to relative motion between two atomic nuclei in an effective potential which is an average Coulomb interaction of nuclei and electron clouds. Morse–like potential models just work with a simple one–dimensional three-parameter effective potential, and find many applications in condensed matter, bio–physics, nano science and quantum optics.

In algebraic approach, the Morse potential models would be rewritten in terms of the generators of $SU(2)$. The quantum relation between q-deformed harmonic oscillator and Morse potential has been considered in many literature such as in [10], where then the anharmonic vibrations of Morse potential have been described as the levels of q-deformed harmonic oscillators.

Journal of Physics: Conference Series **627** (2015) 012022 doi:10.1088/1742-6596/627/1/012022

The extended $SU(2)$ model (the q-Morse potential) was also developed to compare with phenomenological Dunhams expansion and experimental data for numbers of diatomic molecules [10, 11].
In this work, deformed algebra under consideration is mathematical object and atomic effective potential is physical model, we use this relation in inverse way: investigate properties of q-deformed harmonic oscillator on the base of the Morse potential.
The potential of harmonic oscillator is parabolic with infinity equal–step levels. We will show that the potential of q-deformed harmonic oscillator might be described as Morse-like anharmonic potential with finite unequal–step levels. The relation between the deformation parameter q and the set of parameters of Morse-like anharmonic potential has been found. Based on this relation, we investigate thermodynamics of q-deformed harmonic oscillator such as the partition function and some derivative properties.

2. Harmonic Oscillator ($q = 1$)

In the term of first quantization representation, the Hamiltonian of harmonic oscillator is

$$H_0 = \frac{p^2}{2m} + \frac{1}{2} m\,\omega^2 x^2, \tag{1}$$

where m and ω are the mass and frequency of oscillator, respectively. $\hat{x}$ and $\hat{p} = -i\hbar \frac{\mathrm{d}}{\mathrm{d}x}$ are the coordinate and momentum operators, which satisfy the commutatiion relation $[\hat{x}, \hat{p}] = -i\hbar$.
In the second quantization representation, the Hamiltonian of harmonic oscillator is rewritten as

$$H_0 = \frac{\hbar\omega}{2}\left(a_0 a_0^\dagger + a_0^\dagger a_0\right). \tag{2}$$

where $a_0^\dagger$ and a_0 are creation and annihilation operators, which satisfy the commutation relation

$$[a_0, a_0^\dagger] = 1. \tag{3}$$

Energy spectrum of harmonic oscillator has the form

$$E_{0n} = \frac{\hbar\omega}{2}(2n+1), \tag{4}$$

where $n = 0, 1, 2, 3, \ldots$ are the integers.
Note that for harmonic oscillator case, potential $V(x) = \frac{1}{2} m\,\omega^2 x^2$ is parabolic, and energy spectrum has infinite equal-step levels as presented in the fig. 1.

3. q-Deformed Harmonic Oscillator

Creation $a^\dagger$ and annihilation a operators of q-deformed harmonic oscillator satisfy the deformed commutation relation with parameter q [8] as

$$\begin{aligned} [a, a^\dagger]_q &= aa^+ - qa^+a \\ &= 1, \end{aligned} \tag{5}$$

where q is deformation parameter taking values in [0, 1].
In the second quantization representation, the Hamiltonian operator of q-deformed harmonic oscillator is

$$H = \frac{\omega}{2}(aa^+ + a^+a), \tag{6}$$

Journal of Physics: Conference Series **627** (2015) 012022 doi:10.1088/1742-6596/627/1/012022

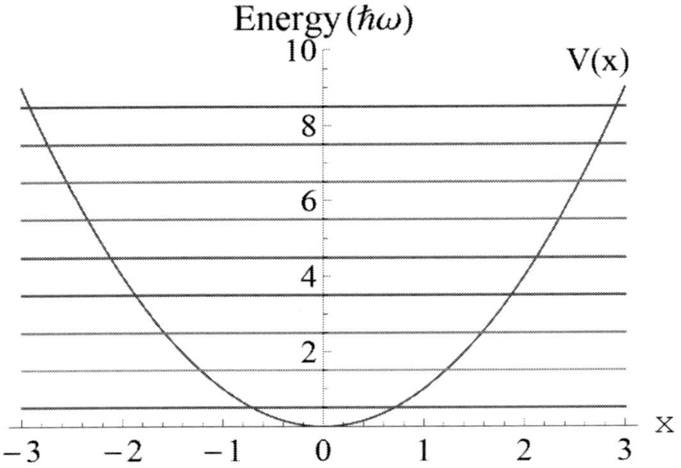

Figure 1. For harmonic oscillator case, the potential is parabolic, and energy spectrum has infinite equal-step levels.

Energy spectrum of q-deformed harmonic oscillator correspondingly takes the form

$$E_n = \frac{\omega}{2}([n] + [n+1]), \tag{7}$$

where $[n] = (1 - q^n) / (1 - q)$ denotes the q-integer which differs from natural numbers.
For small derivation from unity $\varepsilon = 1 - q$, the energy spectrum become quadratic if the higher order contribution $O(\varepsilon^2)$ is neglected

$$E_n = \hbar\omega\left(n + \frac{1}{2} - \frac{n^2}{2}\varepsilon + O\left(\varepsilon^2\right)\right). \tag{8}$$

4. Morse-like Anharmonic Potential

The Morse potential has an important role in describing the interaction of atoms in diatomic, and even polyatomic molecules. Obviously, the Morse potential is anharmonic, and has finite unequal-step level energy spectrum as presented in the fig. 2 where for comparison the harmonic spectrum of corresponding parabolic potential is also plotted. The so-called Morse-like potential is an effective potential with a set of three parameters (D, d, x_e) taken from experiment

$$V_x = D\left(1 - e^{-d(x-x_e)}\right)^2, \tag{9}$$

where x_e is the minimum position, D is the depth of the well, and d is the width of potential. In practice, these parameters are determined from experimental data, D is dissociation energy, and d is restitution constant in the harmonic approximation model as shown in the fig. 2.
The one-dimensional Morse Hamiltonian for oscillator with mass M can be rewritten in term of the generators of $SU(2)$ and reads

$$H_M = \frac{\hbar\omega_M}{2}\left(bb^\dagger + b^\dagger b\right), \tag{10}$$

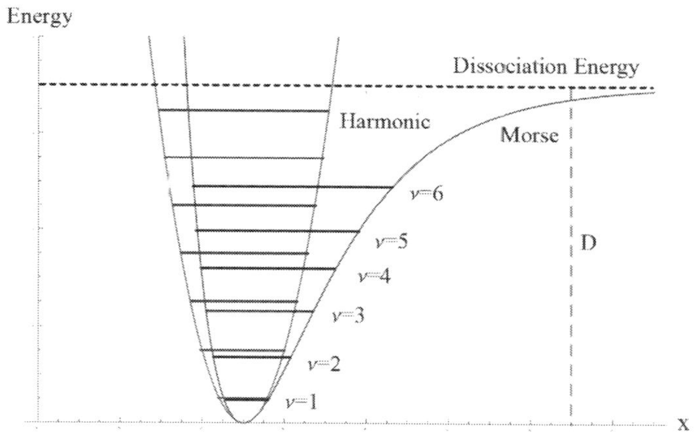

Figure 2. The Morse potential is anharmonic and energy spectrum has finite unequal-step levels, for comparison, the harmonic spectrum of corresponding parabolic potential is also plotted.

where $b^\dagger$ and b are creation and annihilation operators, which satisfy the commutation relations

$$\left[b, b^\dagger\right] = 1 - \frac{2\hat{\nu}}{N}, \tag{11}$$

$$[b, \hat{\nu}] = b, \tag{12}$$

$$\left[b^\dagger, \hat{\nu}\right] = -b^\dagger, \tag{13}$$

where $\hat{\nu}$ is the Morse phonon operator with an eigenvalue. The value of N depends on the depth D and the width d of the Morse potential

$$N = \left(\frac{8Dd^2}{\hbar^2}\right)^{1/2} - 1, \tag{14}$$

and the frequency of Morse oscillator equals to

$$\omega_M = \left(\frac{2D}{md^2}\right)^{1/2}. \tag{15}$$

Energy spectrum of Mosre oscillator is

$$E_{M\hat{\nu}} = \hbar\omega_M\left(\nu + \frac{1}{2} - \frac{\nu^2}{N}\right). \tag{16}$$

where $\nu = 1, 2, 3, ...\nu_{\max}$ is the quanta number. The largest value $\nu_{\max} = [N/2]$, here $[f]$ is the integer part of f. When $N \to \infty$, then $[b, b^+] \to 1$ giving the usual boson commutation relations, we come back to case of harmonic oscillator with parabolic potential and infinity equal-step energy levels.

5. Relation between q-deformed harmonic oscillator and Morse anharmonic vibrations

Compare the energy spectrum of q-deformed harmonic oscillator (8) and those of Mosre oscillator (16), neglecting the higher order contribution $O(\varepsilon^2) = O\left((1-q)^2\right)$, we realize the relations

$$\begin{aligned} \omega_M &\leftrightarrow \omega, \\ \nu &\leftrightarrow n, \\ \varepsilon &\leftrightarrow 1-q. \end{aligned} \tag{17}$$

This is the relation between q-deformed harmonic oscillator and Morse anharmonic vibrations. The changing energy spectrum from linear to quadratic form caused by deformations of commutation relations in the mathematical algebraic approach, while in case of Morse potential in atomic and molecular physics by anharmonic vibrations.

6. Morse potential for deformation model (MPD model)

Using the relation (17), we propose MPD model for investigation properties of q-deformed harmonic oscillator based on the Morse potential. The main idea is that the role of parabolic potential for harmonic oscillator will be replaced by the Morse anharmonic potential for q-deformed harmonic oscillator. In our proposed model, every given value of deformation parameter q in the interval from zero to unity $q \in [0,1]$ can be described by the Morse anharmonic potential with the larges number $n_{\max}$

$$n_{\max}(q) = \left[\frac{1}{(1-q)}\right]. \tag{18}$$

where here $[f]$ is the integer part of f. We note here the well–defined one-to-one corresponding between q and $n_{\max}$. The values of largest number $n_{\max}$ depending on deformation parameter q are plotted in the fig. 3.

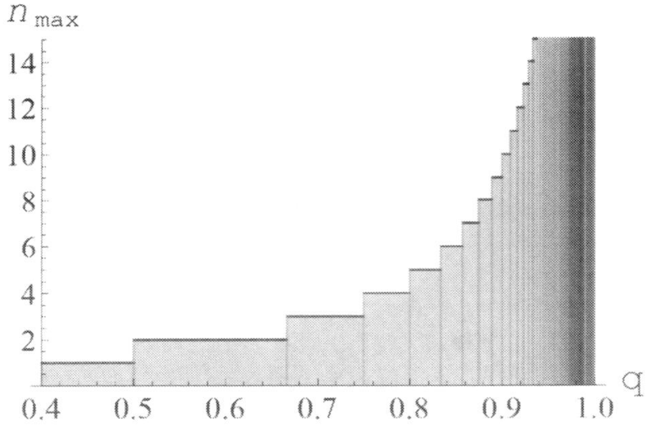

Figure 3. The values of largest number depend on deformation parameter q.

We could see that the total level number rapidly grown up when q tends to unity. In the limit of weak deformation $q \to 1$, $n_{\max} \to \infty$. In contrast, when q tends to 0.5 only one level can be exist. In strong deformation limit $q \to 1/2$, $n_{\max} \to 1$. Consequently, the actual working range of deformation parameter q is not in whole range $q \in [0,1]$, but only in more narrow half range

Journal of Physics: Conference Series **627** (2015) 012022 doi:10.1088/1742-6596/627/1/012022

q	$0 < q < 0.5$	$0.5 < q < 0.66$	$0.66 < q < 0.75$	$0.75 < q < 0.8$	...	$q \to 1$
$n_{\max}$	0	1	2	3	...	$n_{\max} \to \infty$

Table 1. Some first values of deformation parameter q and $n_{\max}$ largest number.

$0.5 < q < 1$.

With a given value of deformation parameter q, the energy spectrum of deformed harmonic oscillator is well–defined

$$E_q(n) = \hbar\omega\left(n + \frac{1}{2} - \frac{n^2}{2\,n_{\max}}\right), \tag{19}$$

where $n = 1, 2, 3, ..., n_{\max}$.

Energy spectrum of deformed harmonic oscillator is well–defined and now has finite unequal–step form. In the fig. 4, four lowest energy levels ($n = 1, 2, 3, 4$) and two higher levels ($n = 10, 20$) of deformed harmonic oscillator depending on deformation parameter q are showed.

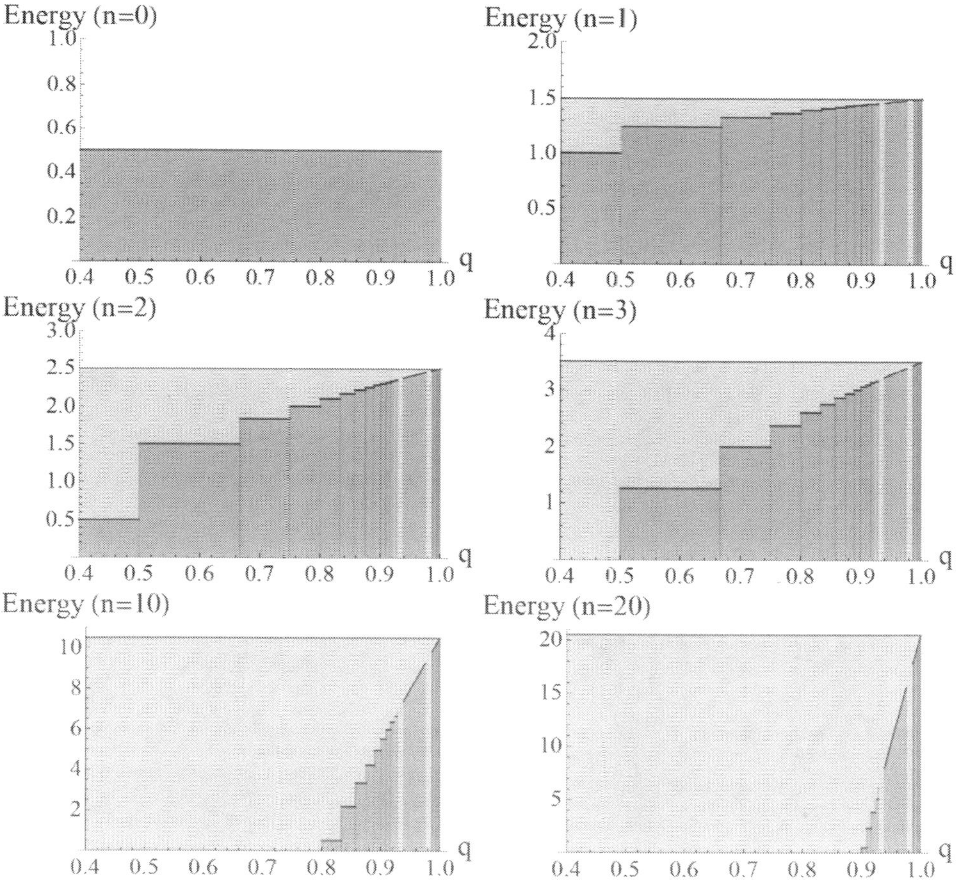

Figure 4. Four lowest energy levels ($n = 1, 2, 3, 4$) and two higher levels ($n = 10, 20$) of deformed harmonic oscillator depending on deformation parameter q.

Journal of Physics: Conference Series **627** (2015) 012022 doi:10.1088/1742-6596/627/1/012022

Higher levels can be occurred when deformation parameter q is very closed to one. The lowest level $n = 1$ occurs if $q = 0.5$, below that point $q < 0.5$ we remain only level $n = 0$ with vacuum oscillation energy $\hbar\omega/2$.
Energy spectrum of deformed harmonic oscillator depends on deformation parameter q and largest number $n_{\max}$ are expressed in the fig. 5. As shown in table 1, begin at $q = 0.5$ we have

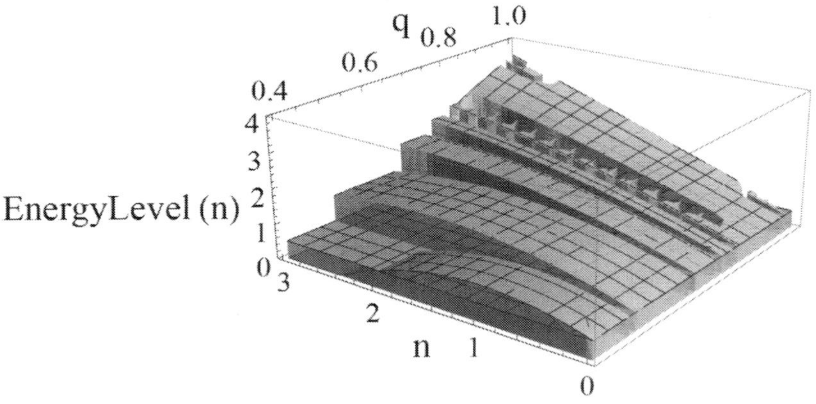

Figure 5. Energy spectrum of q-deformed harmonic oscillator and largest number $n_{\max}$.

one level, at $q = 0.66$ two levels, at $q = 0.75$ three levels, $q = 0.8$ four levels ... and number of energy levels rapidly grown up when q tends to one. In the weak deformation limit $q \to 1$, $n_{\max} \to \infty$ we get back the case of harmonic oscillator with infinite equal-step levels.

7. Deformed partition function

We define the deformed partition function Z_q of deformed harmonic oscillator depending on deformation parameter q as follows

$$Z_q = \sum_{n=0}^{n_{\max}} \exp\left(-\frac{E_q(n)}{k_B T}\right) \tag{20}$$

where k_B is the Boltzmann constant.
With a given value of deformation parameter q, the corresponding values of largest number $n_{\max}$ and energy spectrum $E_q(n)$ of deformed harmonic oscillator are well–defined so deformed partition function Z_q also is well–defined by expression (20).
Using the correspondence between q and $n_{\max}$, the values of deformed partition function Z_q depend on temperature T for some first low lying levels $n_{\max}$= 0, 1, 2, 4, 6 and one high level $n_{\max}$=30 are presented in the fig. 6.
If the given value of deformation parameter q in the range $0 < q < 0.5$ then $n_{\max} = 0$, only one level of vacuum oscillation remains and its partition function is a constant with temperature $Z_q = 1$ as in fig. 6. For large values of q closed to one $0 < q < 1$, $n_{\max}$ might have a large value, so we can approximate take the sum analytically. By doing that we found an explicit expression of Z_q, and it rapidly tends to the partition function of harmonic oscillator when $q \to 1$. For medium q values $0.5 < q < 0.9$, the values of $n_{\max}$ are in the range $1 < n_{\max} < 9$ so the analytical summation is impossible, and we can treat the problem numerically only.
From the deformed partition function Z_q we can derive other related thermodynamic deformed functions such as deformed mean energy, deformed specific heat, the mean number of deformed boson and deformed statistic functions.

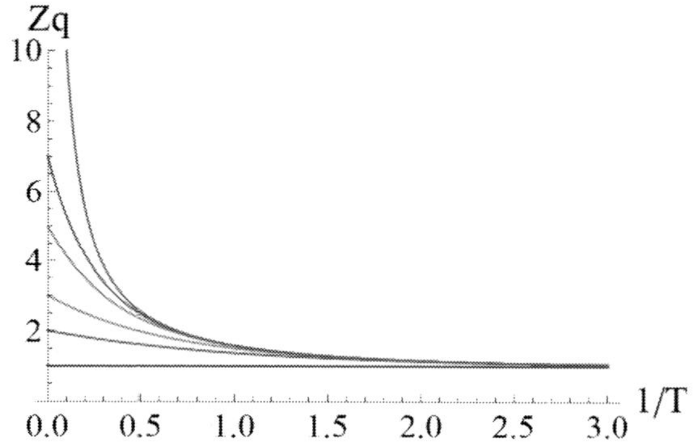

Figure 6. Values of deformed partition function Z_q depend on temperature T for some first low lying levels $n_{\max}$= 0, 1, 2, 4, 6 and one high level $n_{\max} = 30$.

8. Discussion

The main results of this work are obtained by using the relation (17). We proposed a new Morse potential for deformation MPD model for investigating properties of q-deformed harmonic oscillator via the Morse potential. The main idea is that the role of parabolic potential for harmonic oscillator can be replaced by the Morse anharmonic potential for q-deformed harmonic oscillator, so that mathematical deform properties can be described and understood in the language of physical anharmonic behaviours.

In our proposed MPD model, q-deformed harmonic oscillator can be described by the corresponding to Morse anharmonic potential. With every given value of deformation parameter q in the interval $q \in [0, 1]$, we can explicitly define the largest number $n_{\max}$ and the deformed energy spectrum $E_n(q)$ of q-deformed harmonic oscillator.

We have also shown that with a given value of deformation parameter q, the deformed partition function Z_q is also well-defined.

Some explicit properties of larges number $n_{\max}$, deformed energy spectrum $E_q(n)$, and deformed partition function Z_q depending on deformation parameter q are studied for further investigation of other related thermodynamic deformed functions and applications.

References

[1] V. V. Eremin, A. A. Meldianov 2008 *arXiv:0810.1967v1*.
[2] S. S. Mirzrahi, J. P. Camargo, V. V. Dodonov 2004 *Int. J. Phys* A **37**, 3707.
[3] S. Abe 2009 *Int. Phys. Rev* E **79** 041116.
[4] A. Lavagno, P. P. Swamy 2010 *Int. Physica* A **389**, 993.
[5] A. Algin, M. Senay 2012 *Int. Phys. Rev* E **85**, 041123.
[6] F. M. Andrade, E. O. Silva 2013 *Int. Phys. Lett* B **719**, 467.
[7] V. M. Tkachuck 2013 *Int. Phys. Rev* A **86**, 062112.
[8] M. Angelova, A. Franck 2005 *Int. Physics of automic nuclei*, **68**,1689-1697.
[9] S. H. Dong, R. Lemus, and A. Frank 2002 *Int. J. Quant. Phys.* **86** 433.
[10] M. Angelova, V. Dobrev, A. Franck 2004 *Eu. Phys. Jour.* **D31** 27.
[11] I.G. Kaplan 2003 *Handbook of Molecular Physics and Quantum Chemistry, Wiley.*
[12] E. F. Lima and J. E. M. Hornos 2005 *J. Phys. B: At. Mol. Opt.Phys.* **38**, 815.

Influence of a strong electromagnetic wave (laser radiation) on the Hall effect in a cylindrical quantum wires with infinitely high potential

Nguyen Thu Huong[1], **Nguyen Duc Luong**[2], **Nguyen Quang Bau**[2]
[1]Academy of Air Defense and Air Force , Son Tay, Hanoi, Vietnam
[2] Department of physics, College of Nature Science, Hanoi National University, No. 334 Nguyen Trai St., Ha Noi, Vietnam
E-mail: huong146314@gmail.com

Abstract. Based on the quantum kinetic equation for electrons, we theoretically study the influence of a Strong Electromagnetic Wave (EMW) on the Hall effect in a cylindrical quantum wire with infinitely high potential ($V(\overrightarrow{r}) = 0$) inside the wire and $V(\vec{r}) = \infty$ elsewhere subjected to a crossed dc electric field $\vec{E}_1 = (0, 0, E_1)$ and magnetic field $\vec{B} = (0, B, 0)$ in the presence of a strong EMW (laser radiation) characterized by electric field $\vec{E} = (0, 0, E_0 \sin \Omega t)$ (where E_0 and Ω are amplitude and frequency of EMW, respectively). We obtain the analytic expressions for the components σ_{zz} and σ_{xz} of Hall conductivity as well as Hall coefficient with the dependence on B, Ω. The results are numerically evaluated and graphed for GaAs/GaAsAl quantum wire to show clearly the dependence of Hall conductivity and Hall coefficient on above parameters.

1. Introduction

Recently, there has been considerable interest in the behavior of low dimensional systems, in particular of one dimensional electron gas systems such as cylindrical quantum wire, rectangular quantum wire the confinement of electrons in these systems considerably enhances the electron mobility and lead to their unusual behaviors under external stimuli. The Hall effect in bulk semiconductors under the influence of EMWs has been studied in much details [1-5]. In Refs. [1,2] the one magnetoresistance was calculated when the nonlinear semiconductors ares subjected to a magnetic field and an EMW with now frequency, the nonlinearity is resulted from the nonparabolicity of distribution functions of carriers. In Refs [3,4], the magnetoresistance was derived in the presence of a strong EMW for two cases: the magnetic field vector and the electric field vector of the EMW are perpendicular [3], and are parallel [4]. The existence of the odd magnetoresistance was explained by the influence of the strong EMWs on the probability of collision, i.e., the collision integral depends on the amplitude and frequency of the Electromagnetic Waves. This problem was also studied in the presence of both low and high frequency of the EMWs [5]. Throughout these problem, the quantum kinetic equation method have been seen as a powerful tool. In low dimensional semiconductors systems, the carrier confinement leads to unusual behaviors in comparison with bulk semiconductors under external stimuli. So, in recent works, we have used this method to study the influence of an intense EMW on the Hall effect in rectangular and parabolic quantum wells [6,7]. In one dimensional (1D) electron systems, the Hall effect has been studied in many aspects. However, most of

Journal of Physics: Conference Series **627** (2015) 012023 doi:10.1088/1742-6596/627/1/012023

the previous works only considered the cases an EMW was absent and at temperatures that electron - electron anh electron impurity interaction were dominant. In this work, we have used the quantum kinetic equation method to study the influence of a high frequency EMW on the Hall coefficient (HC) in quantum wires. The main purpose of this work is to make a comparison between our calculation and other experiments and theories. Specially we investigate the influence of an EMW on the effect by comparing dependencies of the magnetoresistance and the HC between the absence and the presence of an EMW. The paper is organized as follow. In the next section, we describe the simple model of a Cylindrical Quantum Wires and present briefly the basic formulas for the calculation. Numerical results and discussion are given in Sec. 3. Finally, remarks and conclusions are show briefly in Sec. 4.

2. Hall effect in a cylindrical quantum wires with infinitely high potential in the presence of a laser radiation

2.1. Hamiltonian for electron - phonon interacting system in a cylindrical quantum wires with Infinitely High Potential

We consider a cylindrical quantum wire of the radius R and the length L with the infinite confined potential: $V(\vec{r}) = 0$ inside the wire and $V(\vec{r}) = \infty$ elsewhere subjected to a crossed dc electric field $\vec{E}_1 = (0, 0, E_1)$ and magnetic field $\vec{B} = (0, B, 0)$ in the presence of a strong EMW (laser radiation) characterized by electric field $\vec{E} = (0, 0, E_0 \sin \Omega t)$. Hamiltonian for electron - phonon interacting system in external field can be written as:

$$H = \sum_{n,l,\vec{k}} \varepsilon_{n,l}\left(\vec{k} - \frac{e}{\hbar c}\vec{A}(t)\right) a^{+}_{n,l,\vec{k}} a_{n,l,\vec{k}} + \sum_{\vec{q}} \hbar\omega_{\vec{q}} b^{+}_{\vec{q}} b_{\vec{q}} + \sum C_{n,l,n',l'}(\vec{q}) a^{+}_{n,l,\vec{k}+\vec{q}} a_{n',l',\vec{k}} (b_{\vec{q}} + b^{+}_{-\vec{q}}). \tag{1}$$

where $a^{+}_{n,l,\vec{k}}$ and $a_{n,l,\vec{k}}$ ($b^{+}_{\vec{q}}$ and $b_{\vec{q}}$) are the creation and annihilation operators of electron (optical phonon); $\vec{k} = (0, 0, k_z)$ is the electron wave momentum (along the wire's axis: z axis); $\vec{q}$s the phonon wave vector; $\omega_{\vec{q}}$are optical phonon frequency; $C_{n,l,n',l'}(\vec{q}) = C_{\vec{q}} I_{n,l,n',l'}(\vec{q})$ is the electron optical phonon interaction coefficient; $\left|C_{\vec{q}}\right|^2 = \frac{e^2\hbar\varpi_{\vec{q}}}{2Vk_o q^2}(\frac{1}{\chi_\infty} - \frac{1}{\chi_o})$. Here $V, k_o, \chi_o, \chi_\infty$ are the volume, the electronic constant, the static dielectric constant, and the high frequency dielectric constant, respectively. The electron form factor $I_{n,l,n',l'}(\vec{q})$ can be written as: $I^2_{n,l,n',l'}(\vec{q}) = \left\langle n', l', \vec{k}' \left|e^{i\vec{q}\vec{r}}\right| n, l, \vec{k} \right\rangle$; $\vec{A}(t) = \frac{1}{\Omega} E_0 \cos(\Omega t)$ is the potential vector, depending on the external field. n (l) is azimuthal (radial) quantum number;

Energy Spectrum:

$$\varepsilon_{n,l}(k) = \frac{\hbar^2 k_x^2}{2m^*} + \hbar\omega_c(N_p + \frac{n}{2} + \frac{l}{2} + \frac{1}{2}) - \frac{1}{2m^*}\left(\frac{\mathrm{eE_1}}{\omega_c}\right)^2. \tag{1'}$$

2.2. Quantum kinetic equation for confined electrons in a cylindrical quantum wires with Infinitely High Potential

From Hamiltonian for electron - phonon interacting system in a cylindrical quantum wires with Infinitely High Potential. After several operator calculations, we obtain a set of coupled quantum kinetic equations in which the equation for acoustic phonon can be written as:

$$\frac{f_{n,l}-f_{n',l'}}{\tau_{n,l}}+\left(e\vec{E}_1+\omega_c\left[\vec{k}_x\Lambda\vec{h}\right]\right)\frac{\partial f_{n,l}}{\hbar\partial\vec{k}_x}+\frac{\hbar\vec{k}_x}{m}\frac{\partial f_{n,l}}{\partial\vec{r}}$$
$$=\frac{2\pi}{\hbar^2}\sum_{n,l,n',l',\vec{k}}\left|C_{n,l,n',l'}(\vec{q})\right|^2\frac{\lambda^2}{4\Omega^2}\times$$
$$\times\sum_{\vec{k}}\left[f_{n,l}(\vec{k}-\vec{q})+f_{n',l'}(\vec{k})\right]\delta\left(\varepsilon_{n',l'}(\vec{k})-\varepsilon_{n,l}(\vec{k}-\vec{q})-\hbar\omega_{\vec{q}}-\hbar\Omega\right), \quad (2)$$

where τ is the recovery time of the electron momentum, $f_{n,l}$ is electron distribution, k_B is Boltzainn constant,

$$f_{n,l}=f_o\exp[-\frac{1}{k_BT}\varepsilon_{n,l}(\vec{k})],$$
$$f_{n',l'}=f_o\exp[-\frac{1}{k_BT}\varepsilon_{n',l'}(\vec{k})].$$

2.3. Analytical expressions for the Hall conductivity tensor

We obtain the expression for the conductivity tensor after some manipulation:

$$\sigma_{ij}=\frac{ea\tau}{(1+\omega_c^2\tau^2)}(\delta_{ij}-\omega_c\tau\varepsilon_{ijk}h_k+\omega_c^2\tau^2h_ih_j)$$
$$+\frac{\Gamma_{\vec{q}}}{m}\frac{\tau^2}{(1+\omega_c^2\tau^2)^2}(1-\omega_c^2\tau^2)\delta_{ij}+3(\omega_c\tau^2+\omega_c^4\tau^4)h_ih_j-\omega_c\tau\varepsilon_{ijk}h_k. \quad (3)$$

In this calculation it is assumed $\vec{E}\parallel Oz;\vec{E}=(0,0,E);\vec{B}\parallel Oy;\vec{B}=(0,B,0)$

$$\sigma_{zz}=\frac{\tau}{(1+\omega_c^2\tau^2)}\left[ea+\frac{b}{m}\frac{\tau}{1+\omega_c^2\tau^2}\left(1-\omega_c^2\tau^2\right)\right], \quad (4)$$
$$\sigma_{zx}=\frac{-\omega_c\tau}{(1+\omega_c^2\tau^2)}\left[\frac{ea}{\omega_c}+\frac{b}{m}\frac{\tau^2}{1+\omega_c^2\tau^2}\right], \quad (5)$$

where

$$a=\frac{L}{2\pi}\frac{e\beta\hbar}{m^2}\frac{\tau_o}{1+\omega_c^2\tau_o^2}\exp\left\{\beta\left[\varepsilon_F-\hbar\omega_c(N_p+\frac{n}{2}+\frac{l}{2}+\frac{1}{2})+\frac{e^2E_1^2}{2m\omega_c^2}\right]\right\}\left(\frac{2m}{\beta\hbar^2}\right)^{3/2}\frac{\sqrt{\pi}}{2}, \quad (6)$$
$$b=\sum_{n,l,n',l'}\left|C_{\vec{q}}\right|^2\left|I_{n,l,n',l'}(\vec{q})\right|^2\frac{\lambda^2}{4\Omega^2}\Gamma_{\vec{q}}, \quad (7)$$
$$\Gamma_{\vec{q}}=f_o\frac{Lm^*}{2\pi q\hbar^2}\exp\left\{-\frac{\beta m^*}{2\hbar^2q^2}A^2-\beta A_2\right\}\left(\exp\left[\beta\hbar\left(\omega_{\vec{q}}+\Omega\right)\right]-1\right).$$

And the coefficient Hall is:

$$R_H=-\frac{1}{B}\rho\frac{\sigma_{zx}}{\sigma_{zx}^2+\sigma_{zz}^2}.$$

where $\sigma_{zz};\sigma_{zx}$ are given by Eq. (3)

L is length quantum wire , ρ is the density of quantum wire,

with

$A=\frac{\hbar^2q^2}{2m^*}-\hbar\omega_c(\frac{n'-n+l'+l}{2})+\hbar\omega_{\vec{q}}+\hbar\Omega,$

$A_2=\hbar\omega_c(N_p+\frac{n'+l'+1}{2})-\frac{1}{2m^*}(\frac{eE_1}{\omega_c})^2.$

3. Numerical results and discussions

3.1. The dependence of the Hall coefficient with frequency electromagnetic wave

Survey dependence of the Hall coefficient with frequency electromagnetic waves when the magnetic field changes: $B = 4T; B = 4.4T; B = 4.6T$, length quantum wires $L = 90.10^{-8}m$, $n = 1, n' = 1, l = 1, l' = 1$, radius of the quantum wire $r = 8,77.10^{-9}m$. In Figure 1, we show the dependence of the Hall coefficient with frequency electromagnetic: Initially, as the frequency rises, the Hall coefficient increase before reaching a peak at certain frequency. Then, it falls sharply. And if the electromagnetic wave frequency keep increasing, the Hall coefficient will remain constant. At these values from different schools, different shape figures. There is no difference between the maximum value of the Hall coefficients.

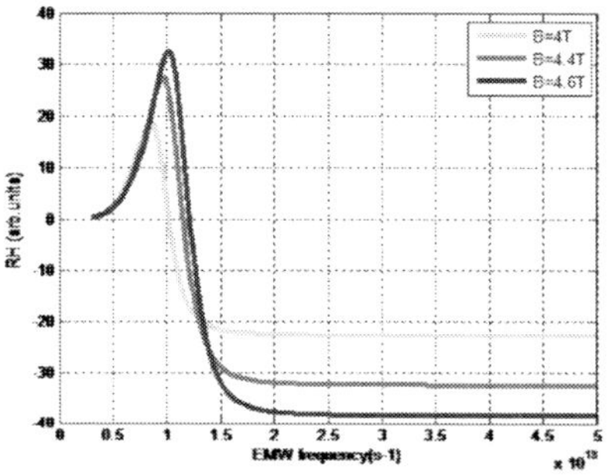

Figure 1. The dependence of the Hall coefficient with frequency electromagnectic wave.

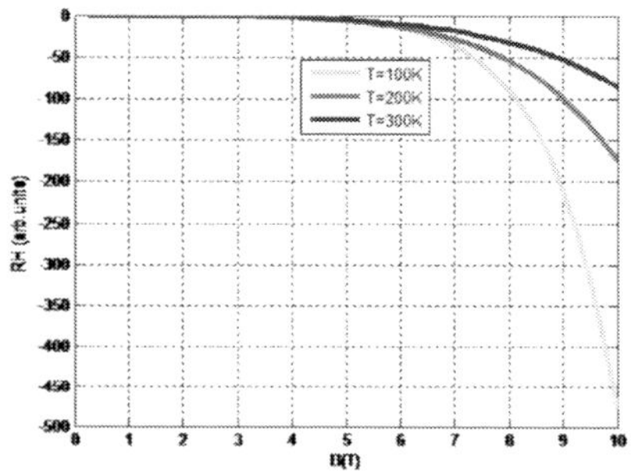

Figure 2. The dependence of the Hall coefficient under a magnetic field $\vec{B}$

3.2. The dependence of the Hall coefficient under a magnetic field $\vec{B}$

Survey dependence of the Hall coefficient on the magnetic field of the system when the temperature changes: $T = 100K, T = 200K, T = 300K$; length quantum wires $L = 90.10^{-8}m$, $n = 1, n' = 1, l = 1, l' = 1$, radius of the quantum wire $r = 8,77.10^{-9}m$. In Figure 2 shows the dependence of the Hall coefficient under a magnetic field $\vec{B}$. Initially, the Hall coefficient has the saturation value. When the magnetic field increases, it decreases . At the low temperatures, the Hall coefficient decreased more rapidly. At a value of the magnetic field ($B = 9T$), the Hall coefficient values corresponding to the low temperature ($100K$) is smaller than those corresponding to the high temperature ($300K$).

4. Conclusion

In this work, we have studied the influence of laser radiation on the Hall effect in a cylindrical quantum wires with infinitely High potential subjected to a crossed dc electric and magnetic fields. The electron optical phonon interaction is taken into account. We obtain the expressions of Hall conductivity as well as Hall coefficient. The analytical results are numerically evaluated and plotted for a specific quantum wires GaAs/AlGaAs to show clearly the dependence of Hall conductivity on the magnetic field, when the magnetic field is small, the Hall conductivity has saturation value. When the magnetic field continues to increase, the Hall conductivity will decrease. The dependence of Hall conductivity with the frequency of electromagnetic waves: Initially, as the frequency rises, the Hall coefficient increase before reaching a peak at certain frequency. Then, it falls sharply, and then fell sharply. The most important thing is the

Journal of Physics: Conference Series **627** (2015) 012023 doi:10.1088/1742-6596/627/1/012023

appearance of the maximum peaks satisfying the (magnetophonon resonance) MPR condition optically detected MPR condition.

References

[1] Epshtein, E. M 1976 *Sov, Phys. Semicond.* **10** 1414.
[2] Epshtein, E. M 1976 *Sov. J. Theor. Phys. Lett.* **2** 234.
[3] Malevich, V. L. and E. M. Epshtein 1976 *Sov. Phys. Solid State. 18* 1286 .
[4] Pankratov, A. A. and E. M. Epshtein 1982 *Sov. Phys. Semicond.* **16** 1689.
[5] Shmelev, G, M., G. I. Tsurkan, and N. H. Shon 1981 *Sov. Phys. Semicond.* **15** 156 .
[6] Bau, N. Q. and Hoi, B. D 2014 *Int. J. Mod. Phys. B* **28** 14 .
[7] Bau, N. Q. and Hoi, B. D 2012 *J. Korean Phys. Soc.* **60** 59 .

IWTCP-2 and NCTP-39 IOP Publishing
Journal of Physics: Conference Series **627** (2015) 012024 doi:10.1088/1742-6596/627/1/012024

General autocatalytic theory and simple model of financial markets

Chu Thuy Anh, Nguyen Tri Lan and Nguyen Ai Viet
Institute of Physics, Vietnam Academy of Science and Technology, 10 Dao Tan, Ba Dinh, Hanoi, Vietnam

E-mail: ctanh@iop.vast.ac.vn

Abstract. The concept of autocatalytic theory has become a powerful tool in understanding evolutionary processes in complex systems. A generalization of autocatalytic theory was assumed by considering that the initial element now is being some distribution instead of a constant value as in traditional theory. This initial condition leads to that the final element might have some distribution too. A simple physics model for financial markets is proposed, using this general autocatalytic theory. Some general behaviours of evolution process and risk moment of a financial market also are investigated in framework of this simple model.

1. Introduction

Autocatalyst system is a well-known concept in physics of complex systems [1], of which elements do not only exist or decay, but might have the possibility of forming new elements or destroying existed elements through recombination of already existing elements. This dynamics is already applied in many biological or sociological evolutions[2][3][4][5].

This paper deals with an economic system, in which each event is considered as a particle interacting with others and possibly producing new particles, therefore the system-environment transforms into a new one. In return, it opens the possibility for chemical reactions that have not been possible before the existence of the new product or the destruction of some existing ones. The concept of the "adjacent possible" [6] or "adjacent probable' '[7] needed to be introduced. This is the set of objects that can possibly become produced within a given time span into the future.

The formation process has been investigated with the cascades of creation and destruction. In this process, element exists in two states 0/1: existing or not existing. A set of elements is formed via an interaction matrix. An element can change its state depending on the states of the elements it is connected to.

This process could be used to investigate financial crisis period. Some information or behaviour could lead to the fact that investigators buy or sell their capitals. If most of investigators buy or sell the same capital then the market is forced to unstable state and there might create a financial crisis.

2. Autocatalyst system

In an evolution, each event i presents in a system with a frequency x_i - normalized abundance of the element i, such that $\sum x_i = 1$, n elements exist, if $x_i > 0$, and does not exist if $x_i = 0$.

Journal of Physics: Conference Series **627** (2015) 012024 doi:10.1088/1742-6596/627/1/012024

Interaction matrix α is used to express inter-dependence between pairs of elements. If the combination of elements j and k are possible to form element i , then $\alpha_{ijk} = 1$, otherwise, $\alpha_{ijk} = 0$. Tensor α indicates the productive sets.

The dynamics of x_i is

$$\frac{d}{dt}x_i = \sum_{j,k} \alpha_{ijk} x_j x_k - x_i \sum_{lj,k} \alpha_{ljk} x_j x_k. \tag{1}$$

In general cases of the catalytic network, n elements are necessary to form a new one instead of two elements, then $j = (j_1, j_2, ..., j_n)$

The relative diversity of the system $a_t \in [0, 1]$ such that $a_t d$ is the number of existing elements at time t. Consider the productive sets in α, that $\alpha_{ijk} = 1$. Only the number r is known, such that rd is the number of such productive sets occurring in the system and that the entries 1 and 0 in α are distributed completely randomly. So r and n are system's characterized variables.

At time $t = 0$, there exist some initial element $a_0 d$, then at time $t \neq 0$ the produced elements are

$$rd \begin{pmatrix} a_t d \\ n \end{pmatrix} \begin{pmatrix} d \\ n \end{pmatrix}^{-1} \cong rd a_t^n. \tag{2}$$

rda_{t-1}^n those elements already have been produced in the preceding time at $t - 1$ and are certainly contained in a_t, only a fraction $1 - a_t$ of remaining elements were not be present in a_t and truly be newly created. Then

$$\begin{aligned} a_{t+1} &= a_t + \triangle a_t, \\ \triangle a_t &= r(1 - a_t)(a_t^n - a_{t-1}^n). \end{aligned} \tag{3}$$

As a_0 is the initial condition then $a_{-1} \equiv 0$. Define $c_t = \triangle a_{t+1}/\triangle a_t$, for long time limit $c \equiv lim_{t\to\infty} c_t$, then

$$c = nr(1 - a_\infty)a_\infty^{n-1}. \tag{4}$$

$$a_\infty = a_0 \sum_{t=0}^{\infty} c^t = \frac{a_0}{1 - c} \tag{5}$$

The $(n + 1)th$ order equation

$$a_\infty - a_0 = nr(1 - a_\infty)a_\infty^n \tag{6}$$

It depends on n that equation (6) takes different forms as:

- $n = 1$ then $a_\infty - a_0 = r(1 - a_\infty)a_\infty$, there is no phase transition of this system,
- $n \geq 2$ there exits a phase transition of system complexity. For the simplest cases only $n = 2$ is considered in this work,

$$a_0 = a_\infty - 2r(1 - a_\infty)a_\infty^2. \tag{7}$$

At $r \sim 2$, there exists a phase transition.

Those solutions are mathematically identical to description of the real gases, i. e. van der Waals equation of state. By definding $V \equiv a_\infty$; $T \equiv a_0(nra_\infty^{n+1})^{-1}$; $P \equiv a_\infty^{-1} + (nra_\infty^{n+1})^{-1}$, then equation (6) can be mapped onto

$$\left(P - \frac{1}{V^2}\right)V = T. \tag{8}$$

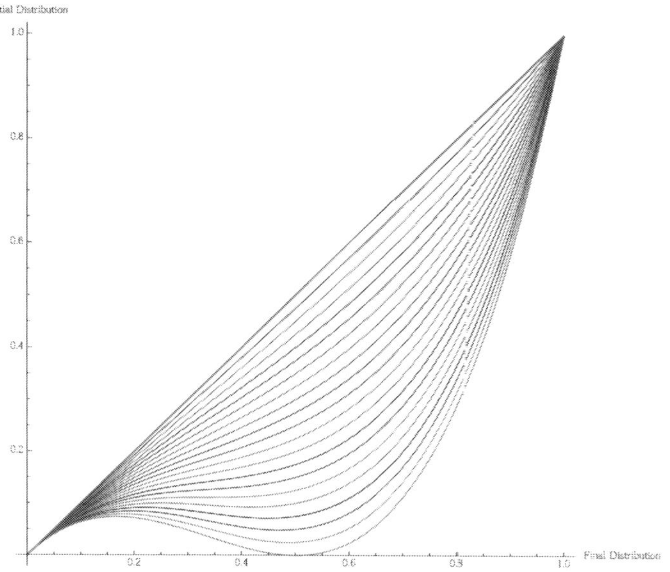

Figure 1. Phase diagram of the creative dynamic for $n = 2$ for various r. A phase transition starts at $r \sim 2$.

This description is a van der Waals gas of point–particles with constant internal pressure. In the context of this work, pressure and temperature are effective ones.

The equation shows the relationship between initial and final states of two elements in autocatalyst process. This equation is equivalent to van der Waals equation of real gas, so the phase transition theory of real gas might be useful to explore the phase transition phenomenon in autocatalyst system, and even in econophysical systems. Using the autocatalyst equation to investigate the transition from Boltzmann distribution to Gaussian distribution, it shows that the autocatalyst can not realize the transition as expected. The initial distribution conserve its form during the autocatalyst process. $\alpha_i j = 1, \alpha_i j = 0$

3. Autocatalyst with a specific complexity of the system

This part presents discussions about two specific cases of the initial complexity of the system, which follows Boltzmann and Gaussian distribution.

In our previous works[8][9], it has been showed that there exists a phase transition of return distributions in econophysics system, such as stocks. A simple model was proposed[10] based on market returns distribution, without any autocatalyst parameter. The model was investigated by using market data of Aluminium Company of America (Alcoa) and DJIA index.

Our previous results[10] showed that there exists the transformation process from Boltzmann-like distribution to Gaussian-like distribution, taking into account time parameter. This dynamic model worked quite good for several emperical data. Then we realised that this process is universal and could be explained by a non-dynamic model with contributions of white noise. Other results [8] showed non-dynamic process, where time parameter was ignored and only background noise contributions were considered. The study with real market data shows that under the influence of background fluctuations, most symmetric distributions tend to Gaussian-like ones before transforming to Gaussian one. These results made an empirical agreement with the central limit theorem.

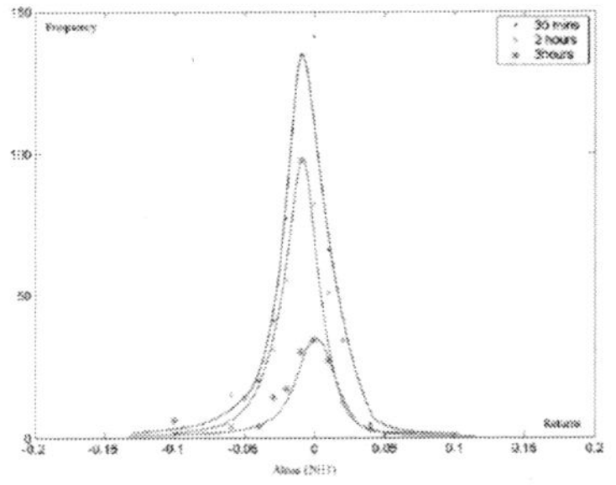

Figure 2. Distribution of Alcoa returns for short time period. The distribution collapsed from Boltzmann-like distribution to Gaussian-like distribution from thirty minutes to one day.

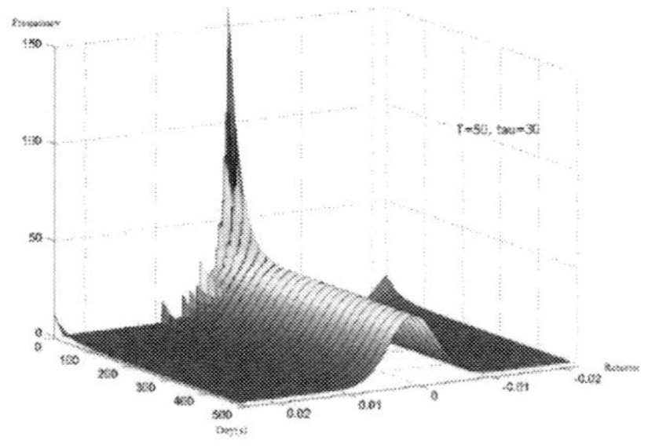

Figure 3. Distribution of DJIA index for medium time period. There is a fast collapse process during the first hundred days from Boltzmann-like distribution to Gaussian-like distribution

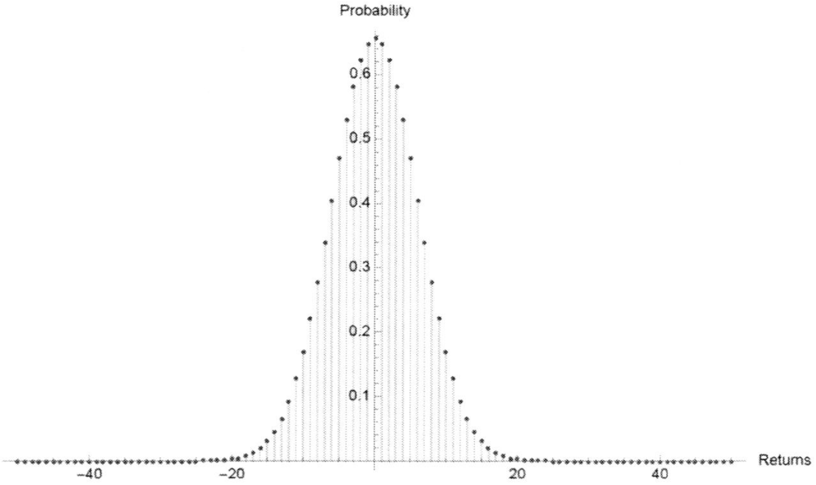

Figure 4. Boltzmann-like distribution collapses to Gaussian-like distribution under noises contribution.

Among various non-dynamic process, autocatalyst could be one of the very first system that could be considered due to analogy between particles interactions and economical interactions[11]. Each agent plays a role as an element in an autocatalyst system, interaction between two agents leads to a new behaviour. Each trading action is considered as an element in this autocatalyst system, represented by its returns. This process would be investigated as an autocatalyst process, with two type of initial returns distribution, Boltzmann and Gaussian distribution. The results are found on Fig. 4 and Fig. 5.

There is no strong emergence, returns distributions keep their type throughout the autocatalyst process. It means that autocatalyst emphasizes the phase transition but phase transition is not a direct result of autocatalyst process.

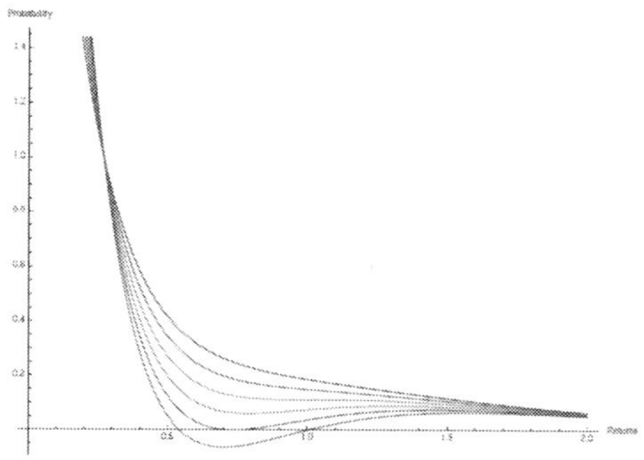

Figure 5. Initial complexity of the system follows Boltzmann distribution. Final distribution of the complexity is Boltzmann still.

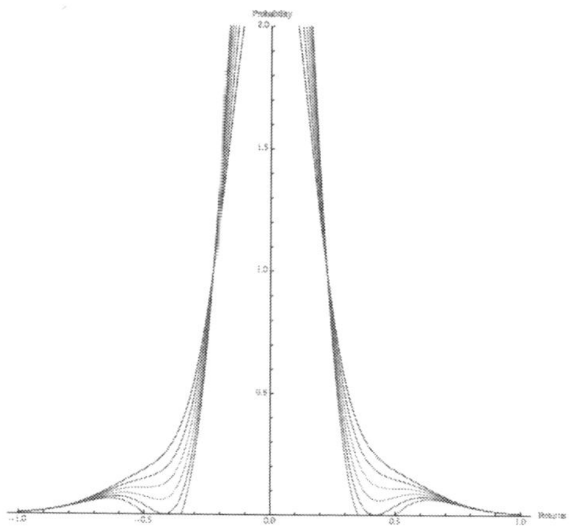

Figure 6. Initial complexity of the system follows Gaussian distribution. Final distribution of the complexity is Gaussian still.

4. Discussion

In this work, the formal equivalence between autocatalyst theory and van der Waals equation has been used to investigate the transition from Boltzmann distribution to Gaussian distribution within the concept of phase transition. As a part of our research in econophysics, returns distribution transformations were considered in order to find out the mechanics of this process. The autocatalyst theory of Hanel *et al.* was used to develop our non-dynamic transformation process, expecting that creation process could lead to a critical transformation in the system. It has been shown that no matter how the initial complexity of the system was, there always exist the phase transition. But it does not like noise contribution process, where no matter what type of initial distribution, general noises lead to Gaussian-like distribution, in this process, autocatalyst emphasizes the phase transition only, but the final distribution of the system does not depend on autocatalyst process.

References

[1] Rudolf Hanel S A K and Thurner S 2005 *Phys. Rev. E* **72**
[2] Schumpeter J A 1911 *Theorie der wirtschaftlichen Entwicklung* (Wien)
[3] Xavier Leoncini M L 2013 *From Hamiltonian Chaos to Complex Systems: A Nonlinear Physics Approach* (Springer)
[4] Debe M K 2012 *Nature* 43–51
[5] Hordijk W and Steel M 2004 *J Theor Biol.* **227** 451–61
[6] Kauffman S 1993 *The origins of order* (Oxford University Press)
[7] Arthur W and Polak W 2006 *Complexity* 23–31
[8] Chu Thuy Anh N T L and Viet N A 2014 *J. Phys.: Conf. Ser.*
[9] Chu Thuy Anh Do Hong Lien N T L and Viet N A 2014 *Communications in Physics* **24**
[10] T A Chu D H L and Viet N A 2013 *Communication in Physics* **23**
[11] Eric Smith D K F 2008 *J. Eco. Dyn. Con.* **32** 7–65

Journal of Physics: Conference Series **627** (2015) 012025 doi:10.1088/1742-6596/627/1/012025

Simple grading model for financial markets

Chu Thuy Anh, Nguyen Tri Lan and Nguyen Ai Viet

Institute of Physics, Vietnam Academy of Science and Technology, 10 Dao Tan, Ba Dinh, Hanoi, Vietnam

E-mail: ctanh@iop.vast.ac.vn

Abstract. A simple way to estimate and grade a financial market by comparison the evolution process and the shape of distribution functions was proposed. In normal working state of financial market, the shape of distribution functions have one-peak form and change from Boltzmann-like to Gaussian-like distributions, while in risk moment might have two-peak form. The grad of financial markets was characterized by overlap area of initial and final distribution functions, and for risk degree by the separation between two shoulders of distribution function. The meaning of Levi tails of distribution and laws of general entropy and information was discussed.

1. Introduction

The transition from an initial distribution, such as Boltzmann distribution appearing in short period data set of econophysics study, to the final Gaussian distribution is a universal phenomenon in several observations in nature, social subjects included. In econophysics, investigation using returns distributions is very common [6, 7, 8], in which Gaussian distribution seemed to be the most used approximation for close-to-stable state [1, 2] of the market. The most observed for disordered state of the market obviously follows Boltzmann distribution. The transition process from disordered state to close-to-stable state of the market has received much attention in theoretical study and also in data mining study. In this framework, a new concept will be introduced to clarify some parts of this process.

There are several parameters affect on the transition from exponential distribution to normal distribution of market returns. In a series of works on econophysical study, few models, dynamic and non-dynamic, were proposed [3, 5][4] to explain the natural insights of this transition, considering an economy system as a basin in contact with environment. The most common input data for these investigation is return index, which can be measure as an internal value of an economy system. Various external values affect on this basin and its transformation process. However, one can not measure all external values, transformation process can not be observed and described in details.

In this work, we will use initial and final distribution without considering the details of transformation process, then will introduce a new concept to describe the changes on the market state. By investigating and comparing initial and final states of markets, information contained in returns distributions can be read out by several ways, reflecting the behaviours that characterize states of market. For a simple consideration, a new measurement tool will be introduced, using returns data to indicate in which state the financial market is, stable or critical ones.

2. Graphical study of overlap

Under several affections such as background noises [4] or dynamic process [3], an initial returns distribution evaluates to some final one. In the transition, the information dissipation from the initial to the final one has been done. However, the information remained in the transition reflects many properties of system and of process that happened along with the transition. In this presentation a simple method based on the geometrical overlap of graphics between initial and final distributions is proposed to estimate and to grade the system under consideration.

As it was mentioned in previous works [3], initial return distribution $P(x)$ of a short term data set is often approximated by Boltzmann form

$$P_B(x) = \frac{1}{2}\lambda e^{-|x|\lambda}, \tag{1}$$

where λ is rate parameter, which is strongly associated with the stock or index liquidity, i.e. the rate parameter is implicitly depended on so-called market temperature. In general, this initial state could be considered as a chaotic state of the market containing a large set of in-time information and causing various scenarios that would happen in next time steps.

Observation of the same data after several internal evolutions, theoretical model and empirical data show that returns rearranged and tends be to Gaussian distribution with variance σ and μ the mean value of x

$$P_G(x) = \frac{1}{\sqrt{2\pi}\sigma}e^{-\frac{(x-\mu)^2}{2\sigma^2}}. \tag{2}$$

In that close-to-stable state, market already follows some deterministic trajectory and market figures are dramatically reduced. Information contained in returns distribution is reduced and more ordered in some aspects. Information transition can be read via examining initial and final returns distributions. It's ordinary to use distribution overlap to study similarity or difference between two given distributions. However, in this work, the overlap is proposed to be a new tool for market crisis measurement reflecting insights of transition process.

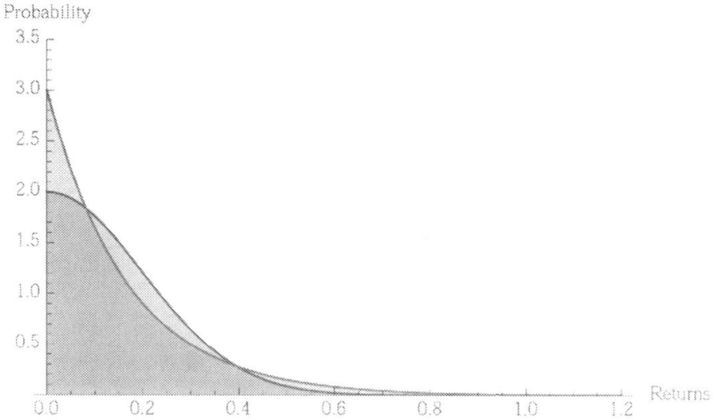

Figure 1. Overlap area of initial and final distribution observed on $x > 0$.

For a big enough data set, distributions are approximatively symmetric, so for the theoretical and analytical study, one can use the continuous version of distributions and the investigation is realized on the positive half axis, $x > 0$ only. Obviously, the equation

$$P_B(x) = P_G(x), \tag{3}$$

has two different solutions, which are

$$x_\pm = \mu + \lambda\sigma^2 \pm \sqrt{2\lambda\mu\sigma^2 + \lambda^2\sigma^4 - 2\sigma^2 \mathrm{Log}\left[\sqrt{\tfrac{\pi}{2}}\lambda\sigma\right]}. \tag{4}$$

Then overlap area of two distributions is determined by expression

$$O_{i,f} = \frac{1}{2}\left(e^{-\lambda x_-} - e^{-\lambda x_+}\right) + \frac{1}{2}\left(\mathrm{Erf}\left[\frac{\mu}{\sqrt{2}\sigma}\right] + \mathrm{Erf}\left[\frac{x_- - \mu}{\sqrt{2}\sigma}\right] + \mathrm{Erfc}\left[\frac{x_+ - \mu}{\sqrt{2}\sigma}\right]\right). \tag{5}$$

It is well known that the overlap area of initial and final distribution contains appearance possibilities of both initial and final states. It could be used to measure the stability of the market.

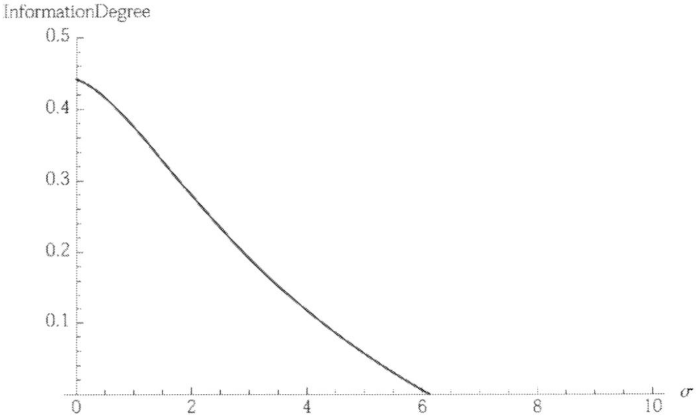

Figure 2. The change of overlap area in σ.

An Information Degree denoted by I is introduced to describe market states.

$$I = \frac{O_{i,f}}{100}. \tag{6}$$

The newly defined information degree might vary from 0 to 100. The 0 degree corresponds to a complete chaotic state, 100 degree corresponds to almost stable state. It's opposite to thermal degree where higher degrees correspond to more disordered states.

In other hand, for a study on the real market data, the discrete version would be employed. In this context, the overlap is numerically determined by following sum

$$O_{i,f} = \frac{1}{2}\Delta x \sum_l \left[\theta\left(P_{f,l} - P_{i,l}\right)P_{i,l} + \theta\left(P_{i,l} - P_{f,l}\right)P_{f,l}\right]. \tag{7}$$

The above expression will be used in our future extensive study on real market data.

3. Conclusion

A simple way to estimate and to grade a financial market by comparison the evolution process and the shape of distribution functions was proposed. The investigation within the short and long period data sets is performed graphically. A new index, Information Degree, has been introduced to determine in which state the market is. Information Degree used in this paper is different from other statistical view on fat tail distribution, where most information is got and analysed for tails part only, but as shown, it could be considered as another measurement tool which is easier to use by economists, especially who work on financial markets.

As Information Degree is of inverse order to thermal degree, we wish to make a connection between the two in order to simplify the tool and to bring more clear physical meaning to our newly proposed approach. Further consideration will take place to define more practical quantities for real financial markets.

References

[1] Victor M. Yakovenko A. Christian Silva, Richard E. Prange. Exponential distribution of financial returns at mesoscopic time lags: a new stylized fact. *Physica A*, (344), 2004.

[2] V. M. Yakovenko A. Dr. Probability distribution of returns in the heston model with stochastic volativity. *Quant. Finance 2*, (443), 2002.

[3] D. H. Lien Chu. T. A and N. A. Viet. Simple model for market returns distribution. *Communications in Physics*, 23(2), 2013.

[4] N. T. Lan Chu T. A. and N. A. Viet. Boltzmann-gaussian transition under specific noise effect. *J. Phys.: Conf. Ser.*, (537), 2014.

[5] N. T. Lan Chu. T. A, D. H. Lien and N. A. Viet. Study of Hanoi and Hochiminh Stock Exchange by Econophysics Methods. *Communications in Physics*, 24(3S2), 2014.

[6] Stauffer D. Phys. rev. e 60 (1999) 6519. *Physica A*, (336):1–5, 2004.

[7] H. Salunkay K. Matia, M. Pal and H.E. Stanley. *Europhys. Lett.*, 66(909), 2004.

[8] V. M. Yakovenko. Econophysics, statistical mechanics approach to. In *Encyclopedia of Complexity and System Science*. Springer, 2007.

Analysis the UV-visible Spectra of Neuroglobin Based on Two-Level Model

T T Thao, C T Anh, N T Lan and N A Viet

Institute of Physics, Vietnam Academy of Science and Technology, 10 Dao Tan, Thu Le, Ba Dinh, Ha Noi, Vietnam

E-mail: ttthao@iop.vast.ac.vn

Abstract. Neuroglobin (Ngb), a novel member of the Globin Group, is recently discovered by Burmester et al. (2000). Its uncertain physiological function makes lots of interest. The existing of a six-coordination heme geometry with proximal and distal histidines directly creates an axis within the heme iron, while the sixth ligand coordination binds to small ligand reversibly. The analysis of UV-visible spectrum of Ngb by the well-known two-level model shows an agreement of the experiment data and theoretical results.

1. Introduction

Hemoglobin (Hb) and myoglbin (Mb) are types of globin family, having been studied for more than four decades. These proteins are characterized by a highly conserved α-helical fold and the ability to reversibly bind dioxygen and other small ligands at the central iron of a prosthetic group (protoheme IX). Neuroglobin (Ngb), namely the oxygen transporters/storage proteins, was recently discovered in the vertebrate globin family [1–3].

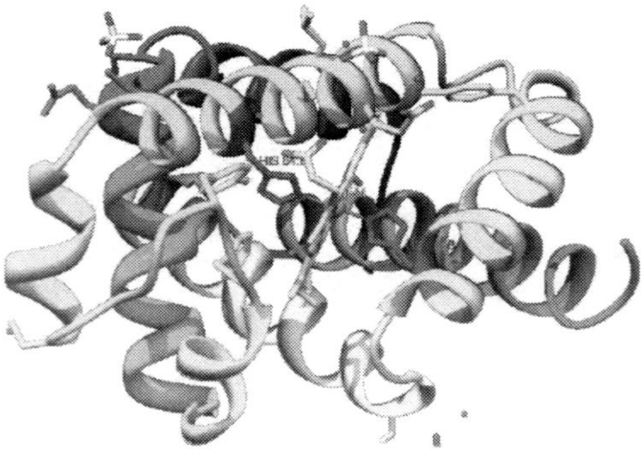

Figure 1. A model of the wild-type human neuroglobin (1OJ6.pdb file), in UCSF Chimera visualization.

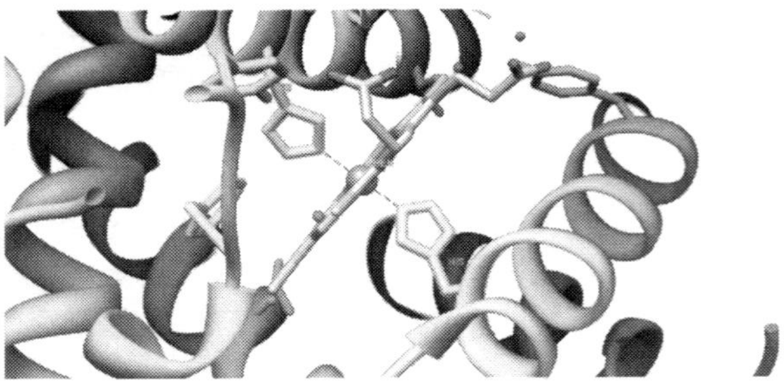

Figure 2. Zoom of active site of Ngb.

Having a single chain of 151 amino acids, Ngb has less than 25% sequence identity with its more prominent cousins Hb and Mb, exhibit a classical three-over-three $\alpha-$ helix sandwich globin fold [3,4]. In the absence of an external ligand, the ferric (Fe^{+3}) and ferrous (Fe^{+2}) heme-iron of Ngb are both hexacoordinated, with histidyl imidazole of proximal His 96 (F8) and distal His 64 (E7) directly bound to the ion metal, whereas Hb and Mb are both pentacoordinated globins with only the proximal histidine bound to the iron. In the presence of external ligand (O_2, CO and NO), only the ferrous form is observed. This feature has been proposed as a novel mechanism for the regulation of ligand affinity in heme proteins [5,6] but its functional significance is a problem of debate.

In Mb and Hb, the Fe^{2+} atom of the heme prosthetic group binds O_2 occurring on the distal side of a pentacoordinated heme, where O_2 establishes a sixth coordination bond to the heme Fe. While, in Ngb, the binding of O_2 is generally stabilized by interaction with distal residues. Often, in vertebrate globins, the main O_2 stabilizing interaction is provided by a hydrogen bond donated by residue HisE7.

Over the years, the bonding of O_2 to the iron-binding site of the heme protein makes a lots of interest [7–10], according to the strong electronic correlation effects associated with its localized Fe 3d electrons. It is well known that these electrons are energetically well-aligned with the π^* acceptor orbitals of CO and O_2, and that the molecules bound conformations seek to maximize intermolecular orbital overlap [11,12]. While the spectra of oxygen binding to hemprotein is quite similarly to each other, the UV-vis spectra of wild-type neuroglobin (WT-Ngb) has also the same their feature. Therefore, in this report, we apply the two-level model to investigate the spectra of WT-Ngb.

2. Material and methods

The material and methods are detailed in the ref [13]. A sample contain WT-Ngb was examined in the spectral range of 280-700 nm, in condition of pH=2.3.

2.1. Crystal field theory for Fe^{2+} complex system

It is know that electron configuration of Fe is $1s^22s^22p^63s^23p^64s^23d^6$. Fe^{2+} has 6 d electrons so its configuration is $1s^22s^22p^63s^23p^63d^6$.

Neuroglobin contains a porphyrin ring with an iron at its center. Two histidines are the axial ligands of the Fe atom (Fig. 2), both in the Fe^{2+} and Fe^{3+} states, the heme is hexacoordinated. However, only ferrous state Fe^{2+} of heme-iron can binds to O_2.

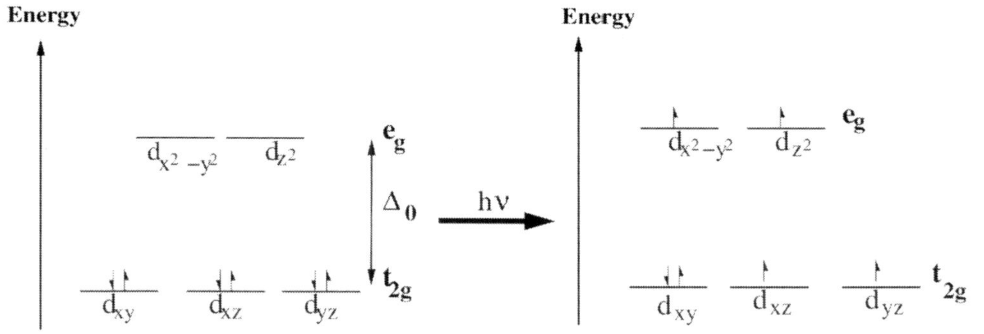

Figure 3. The energy level scheme for ferrous (Fe^{2+}) iron showing the occupation of the 3d orbitals (left) and d-d transitions when the electrons absorb the photons of light (right) [14,15].

Journal of Physics: Conference Series **627** (2015) 012026 doi:10.1088/1742-6596/627/1/012026

Following crystal field theory [14, 15], the Fe^{2+} complex has an octahedral shape. Fe^{2+} of neuroglobin has 6-coordination so its state is in low spin with weak ligand field. For the distorted octahedral geometry, the 5 degeneracy of d-orbitals split in sets of 2 (e_g) and 3 (t_{2g}) obitals (Fig. 3, left)

3 obitals of low energy: d_{xy}, d_{xz}, d_{yz}

2 obitals of high energy: d_{z^2}, $d_{x^2-y^2}$

The splitting between the t_{2g} and e_g orbitals stays constant at Δ_0 regardless of the nature of the distortion.

In the d–d transitions (Fig. 3, right), the electrons in the t_{2g} orbitals of Fe^{+2} ion in octahedral complex absorbs photons of light with energy equal to Δ_0, which causes the electron to move to an empty or singly occupied e_g orbit. The spectra of the absorbance is in UV-vis region. In this work, we apply a simple model to explain the peak of UV-vis absorbance spectra of WT-Ngb.

2.2. *Two-level model*

The figure 3 shows the model of structure of WT-Ngb and the energy level scheme for ferrous (Fe^{2+}) iron.

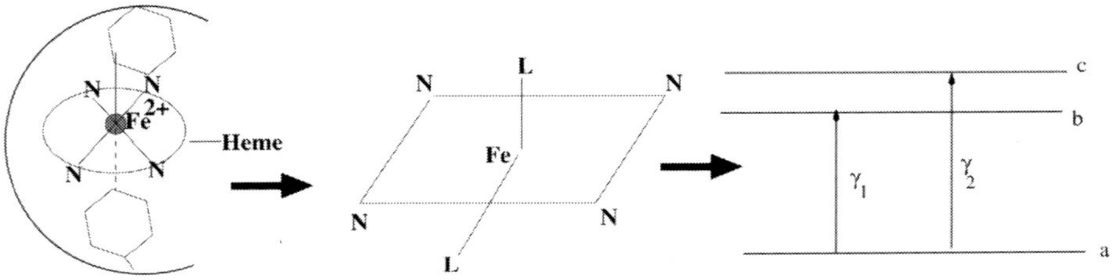

Figure 4. Two-level model diagram of Ngb.

Considering living time of excited state of electron in Fe^{+2} heme complex is much sorter than its living time of ground state, from uncertainty principle we can write for WT-Ngb complex system as follows

$$H = E_0 a^+ a + (E_1 + i\Gamma_1) b^+ b + (E_2 + i\Gamma_2) c^+ c + \gamma_1 (a^+ b + b^+ a) + \gamma_2 (a^+ c + c^+ a). \quad (1)$$

Now we have 2 levels:

- Level 0, ground state energy E_0:
  - $|0\rangle$ is the ground state of Fe^{+2} heme complex
  - a and a^+ are the annihilation and creation operators of Fe^{+2} heme complex in ground state.
- Level 1, excited state energy E_1 and E_2:
  - $|1\rangle$ is the excited state of Fe^{+2} heme complex,
  - b and b^+; c and c^+ are the annihilation and creation operators of Fe^{+2} heme complex in excited states E_1 and E_2, respectively.

This simple two-level model has a set of 4 parameters, which can be taken from experiments:

$\Delta E_1 = E_1 - E_0 = P_1$ and $\Delta E_2 = E_2 - E_0 = P_2$ are the pairing energies of the absorption peaks,

γ is the transition matrix element between the two levels (effective coupling coefficient with photon) characterized by the height of absorption peak,

Γ is the effective width of the excited state (or damping constant) characterized by the broaden of absorption peak from delta form.

So, absorption coefficient

$$A_{abs} \sim g^2 \delta(E_i - E_0 - \hbar\omega + i\Gamma), \tag{2}$$

where ω is photon frequency, $\delta(x + iy)$ is the general delta Dirac function with Gaussian form

$$A_{abs} \sim exp\left[-\left(\frac{E_i - E_0 - \hbar\omega}{\Gamma_i}\right)^2\right] = exp\left[-\left(\frac{P_i - \hbar\omega}{\Gamma_i}\right)^2\right], \tag{3}$$

where $i = 1, 2$ denote the excited states. By fitting theory with experiment data, we get numerical values of the parameters.

3. Results and discussion

In this work, we use the Gaussian functions

$$f(x) = \frac{cf_1}{\Gamma_1\sqrt{2\pi}} exp(-\frac{(x-\mu_1)^2}{\Gamma_1^2}) + \frac{cf_2}{\Gamma_2\sqrt{2\pi}} exp(-\frac{(x-\mu_2)^2}{\Gamma_2^2}). \tag{4}$$

in order to find the position μ and the width Γ of the peaks in the absorption spectra of WT-Ngb.

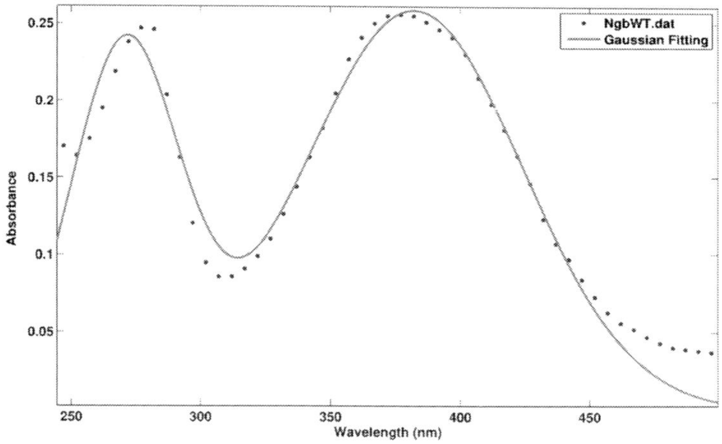

Figure 5. Fitting curve of spectra of WT-Ngb with pH = 2.3. Numerical calculation with ration 1/2: μ_1 =270.7 nm, $\Gamma_1 = 17.47, cf_1 = 17.47, \mu_2 = 382nm, \Gamma_2 = 59.39, cf_2 = 38.52$.

The UV-vis absorption spectra of WT-Ngb (Fig. 5) shows that there are two peaks at positions $\sim 270nm$ and $\sim 382nm$, which displays the same feature of oxyheme protein spectrum. It means that, the heme complex structure of Ngb differs from Hb or Mb one (which presents the typical penta-coordinated geometry). Moreover, we see that two peaks are not similar. Therefore, we suppose that the octahedral geometry of Fe^{2+} heme complex of Ngb is distorted: two bonds of axial distal are not equidistant with the bonds of Fe-N.

There is an agreement between experimental data and numerical calculation. The simple two-level model can be present quantum chemical calculation and can be used for investigating the properties of big molecules.

4. Acknowledgments

We would like to thank Dr. Sophie Bernad for giving us the UV-vis spectra of WT-Ngb.
We acknowledge partial financial support from the Lotus project No.45/2012/HĐ-NĐT.

References

[1] Burmester T et al. 2000 *Nature* **407** 520.
[2] Burmester T et al. 2002 *EMBO Rep.* **3**(12) 1146.
[3] Pesce A, et al. 2003 *Structure* **11** 1087.
[4] Vallone B, et al. 2004 *Proteins Struct Funct Genet*; **56** 82.
[5] Pesce A, et al. 2004 *IUMB Life* **56** 657.
[6] Dewilde S, et al. 2001 *J. Biol. Chem.* **276** 38949.
[7] Goddard W A, et al. 1975 *Proc Natl Acad Sci USA* **72**(6), 2335.
[8] Jensen K P, et al. 2005 *J. Inorg. Biochem.* 45.
[9] Chen H, et al. 2008 *J Am Chem Soc* **130**(44) 14778.
[10] Radon M and Pierloot K 2008 *J Phys Chem A* **112**(46), 11824.
[11] Reed C A, Cheung S K 1977 *Proc Natl Acad Sci USA* **74**(5), 1780.
[12] Vojtechovský J, et al. 1999 *Biophys J* **77**(4), 2153.
[13] Thao T T et al. 2014 *J.Phys.: Conf.Series* **537**, 012006.
[14] Brian N. Figgis, *Ligand Field Theory*, University of Western Australia, Nedlands, Australia and Brookhaven National Laboratory, Upton, NY, USA.
[15] Bethe H 1929. "Splitting of Terms in Crystals". *Ann. Physik* **3**,133-206.

Effects of macromolecular crowding on protein folding

Bui Phuong Thuy[1,2], **Hoang Thi Thu Huong**[1] **and Trinh X Hoang**[1]
[1] Institute of Physics, Vietnam Academy of Science and Technology, 10 Dao Tan, Ba Dinh, Ha Noi, Vietnam
[2] Nam Dinh University of Technology Education, Phu Nghia, Loc Ha, Nam Dinh, Vietnam
E-mail: buiphuongthuyphys@gmail.com

Abstract. Folding of proteins *in vivo* is influenced by the presence of a high concentration of macromolecules in the cytosol. In this study we investigate the effects of macromolecular crowding on folding of proteins by using molecular dynamics simulation method with Langevin equations. The protein is considered in a Go-like model whereas crowders are modeled as soft spheres. In our simulations, protein and crowders are enclosed in a spherical container. It is shown that the crowders enhance the folding transition temperature as well as folding stability of protein. The effect of crowders on the folding free energy is found to agree with Minton's scaled particle theory.

1. Introduction

The cytosol of living cells contains a large number of macromolecules such as proteins, nucleic acids of a volume fraction up to about 40%. Such condition, known as 'macromolecular crowding' [1], significantly changes the properties of molecules in the solution and plays an important role in many cellular processes [2, 3, 4]. For example, macromolecular crowding is known to strengthen the binding affinity between DNA and DNA polymerase in DNA transcription and replication [5], to enhance protein-protein association [6], and to improve the functional efficiency of chaperonins [3]. Macromolecular crowding also has been shown to increase stability and folding rate of proteins [7] and at the same time promote their aggregation [8].

The primary effect of macromolecular crowding is due to the excluded volume occupied by the molecules of the crowd, or crowders, that becomes unavailable to other molecules. In protein folding, fluctuation of the void spaces left by the crowders disfavors the unfolded state more than the folded state. This is the basis of a number of theories based on scaled particle theory (SPT) [9, 10, 11] that allow one to estimate the change in the folding free energy due to a presence of the crowders. Molecular dynamics simulations have confirmed this picture and provided detailed assessments of these theories [12, 13]. The effects of crowding on protein folding have been shown to be essentially the same as those of confinement [11, 14].

In this study, we investigate the effects of macromolecular crowding on folding of protein by molecular dynamics simulation. We consider simple models of protein and crowders that are essentially similar to those considered in Ref. [13] except that we use a close boundary condition instead of a periodic boundary condition.

2. Model and methods

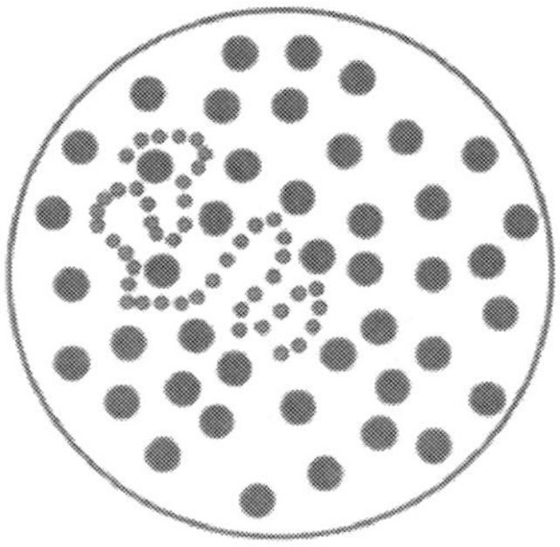

Figure 1. Sketch of a macromolecular crowding model with a protein represented by a chain of amino acids (small dots) and other macromolecules (large dots) enclosed in a spherical container.

Figure 2. The native state of protein G (B1 domain) in a ribbon presentation. The structure has a β-sheet with 4 strands (blue) and an α-helix (red).

2.1. Models of protein and macromolecular crowders

To mimic the cellular environment, we simulate folding of protein in presence of macromolecules modeled as soft spheres (Fig. 1). The protein is considered in a coarse-grained C_α-based Go-like model [15, 16]. Each amino acid is considered as a single bead of radius $R_a = 2.5\,\mathring{A}$ centered at the position of its C_α atom. The potential energy of the protein in a given conformation is given by:

$$E_p(\{\mathbf{r}_i\}) = \sum_{i=1}^{N-1} K_b(r_{i,i+1} - b)^2 + \sum_{i=2}^{N-1} K_\theta(\theta_i - \theta_i^*)^2 + \sum_{n=1,3} \sum_{i=2}^{N-2} K_\phi^{(n)}[1 + \cos(n(\phi_i - \phi_i^*))] + + \sum_{i+3<j} 4\epsilon \left[\left(\frac{\sigma_{ij}}{r_{ij}}\right)^{12} - \left(\frac{\sigma_{ij}}{r_{ij}}\right)^{6} \right] \Delta_{ij} + \sum_{i+3<j} \epsilon \left(\frac{\sigma}{r_{ij}}\right)^{12} (1 - \Delta_{ij}) \, , \quad (1)$$

where N is the number of amino acids; $\mathbf{r}_i$ is the position of the bead i ($i = 1, \ldots, N$); r_{ij} is the distance between beads i and j; θ and ϕ are the bond angle and dihedral angle; n is equal to either 1 or 3; the superscript $*$ corresponds to the native state; Δ_{ij} is equal to 1 if there is a native contact between i and j and equal to 0 otherwise. A native contact is defined if the distance between two residues in the native state is less than 7.5 $\mathring{A}$. The first three terms of Eq. (1) correspond to the bonding potentials, bond angle potentials and dihedral angle potentials, respectively. The last two terms are Lennard-Jones (LJ) potentials for native contacts and repulsive potentials for non-native contacts. Energy is given in units of ϵ, which corresponds to the depth of the LJ potential. The choice of LJ potential is such that the distance between residues in a native contact in the native state corresponds to its minimum, i.e. $\sigma_{ij} = 2^{-1/6} r_{ij}^*$. The parameters chosen for our model are $b = 3.8\mathring{A}$, $\sigma = 2R_a = 5\mathring{A}$, $K_b = 100\,\epsilon\mathring{A}^{-2}$, $K_\theta = 20\,\epsilon(\text{rad})^{-2}$, $K_\phi^{(1)} = -\epsilon$, $K_\phi^{(3)} = -0.5\epsilon$.

Macromolecular crowders are modeled as soft spheres of radius R_c. They repulse each other and the amino acids via a shifted and cut-off LJ potential:

$$U(r') = \begin{cases} 4\epsilon \left[\left(\frac{\sigma}{r'}\right)^{12} - \left(\frac{\sigma}{r'}\right)^{6} \right] + \epsilon & r' \le r_c \\ 0 & r' > r_c . \end{cases} \quad (2)$$

Journal of Physics: Conference Series **627** (2015) 012027 doi:10.1088/1742-6596/627/1/012027

with

$$r' = r - R - R_c + r_c, \tag{3}$$

where r is the center to center distance between the two molecules, R is either the radius of an amino acid (R_a) or the radius of a macromolecule (R_c), and $r_c = 2^{1/6}\sigma$.

We employ a boundary condition in which protein and crowders are enclosed in a spherical container of radius R_{wall} as shown in Fig. 1. The repulsive potential between amino acids and macromolecular crowders with the wall of the container has the same form as given in Eq. (2) but with:

$$r' = R_{\text{wall}} - r - R + r_c, \tag{4}$$

where r is the distance from the center of the amino acid or the crowder to the center of the spherical container. The volume fraction of M identical crowders in a volume V is given by:

$$\phi_c = \frac{(4\pi/3)R_c^3 M}{V}. \tag{5}$$

2.2. *Molecular Dynamics method*

The molecular dynamics (MD) simulation method employed in the present study is based on Langevin equation of motion [16]. For a given temperature T, Langevin equations for a protein of N amino acids and M crowders are given by:

$$m\,\ddot{\mathbf{r}}_i(t) = \mathbf{f}_i(t) - \zeta_a\,\dot{\mathbf{r}}_i(t) + \sqrt{2\zeta_a k_B T}\,\mathbf{g}(t) \qquad (i = 1,\ldots,N) \tag{6}$$

$$m_c\ddot{\mathbf{r}}_j(t) = \mathbf{f}_j(t) - \zeta_c\,\dot{\mathbf{r}}_j(t) + \sqrt{2\zeta_c k_B T}\,\mathbf{g}(t) \qquad (j = 1,\ldots,M) \tag{7}$$

where m and m_c are the masses of an amino acid and a crowder, respectively; $\mathbf{f}_i$ and $\mathbf{f}_j$ are molecular forces; ζ_a and ζ_c are friction coefficients for amino acid and crowder, respectively; $\mathbf{g}$ denotes an "unit" random force whose components are given by a Gaussian white noise of zero mean and unit dispersion.

A Verlet algorithm is developed to numerically integrate the equations. Temperature is given in units of ϵ/k_B, whereas time is measured in units of $\tau = \sqrt{m\sigma^2/\epsilon}$. The integration time step is chosen to be equal to $\Delta t = 0.002\tau$. We assume that the protein and the crowder have the same mass per volume, so that $m_c = m(R_c/R_a)^3$. The friction coefficients for amino acids and crowders are chosen to be $\zeta_a = 5\,m/\tau$ and $\zeta_c = \zeta_a(R_c/R_a)$, so that they follow the Stokes law for viscous friction on spherical particles.

Equilibrium characteristics of the system can be calculated from data of MD simulations and the weighted histogram method [17]. The specific heat of protein is given by:

$$C_v(T) = \frac{\langle E^2\rangle - \langle E\rangle^2}{k_B T^2}, \tag{8}$$

where E the total energy of the protein. The protein folding transition temperature or folding temperature T_f is defined as the temperature of the maximum of the specific heat. For a given conformation one defines the fraction of native contact Q as the number of native contacts in a given conformation divided by the total number of contacts in the native state. At a given temperature T an effective free energy as function of Q can be calculated as:

$$F(Q) = -k_B T \ln P(Q), \tag{9}$$

where $P(Q)$ is the probability to find the protein having native contact fraction between Q and $Q + dQ$.

3. Results and discussion

We study the effects of crowding on folding of the B1 domain of protein G (GB1), a small protein of $N = 56$ amino acids with native state shown in Fig. 2. The radius of the crowder is fixed to be $R_c = 10.2\mathring{A}$, that is equal to the radius of gyration of GB1, and the radius of the container to be $R_{\rm wall} = 106.4\mathring{A}$. The size of the container was chosen so that it can accommodate a fully stretched protein. Extensive simulations are carried out to obtain equilibrium characteristics of the folding transition for different volume fractions of crowders.

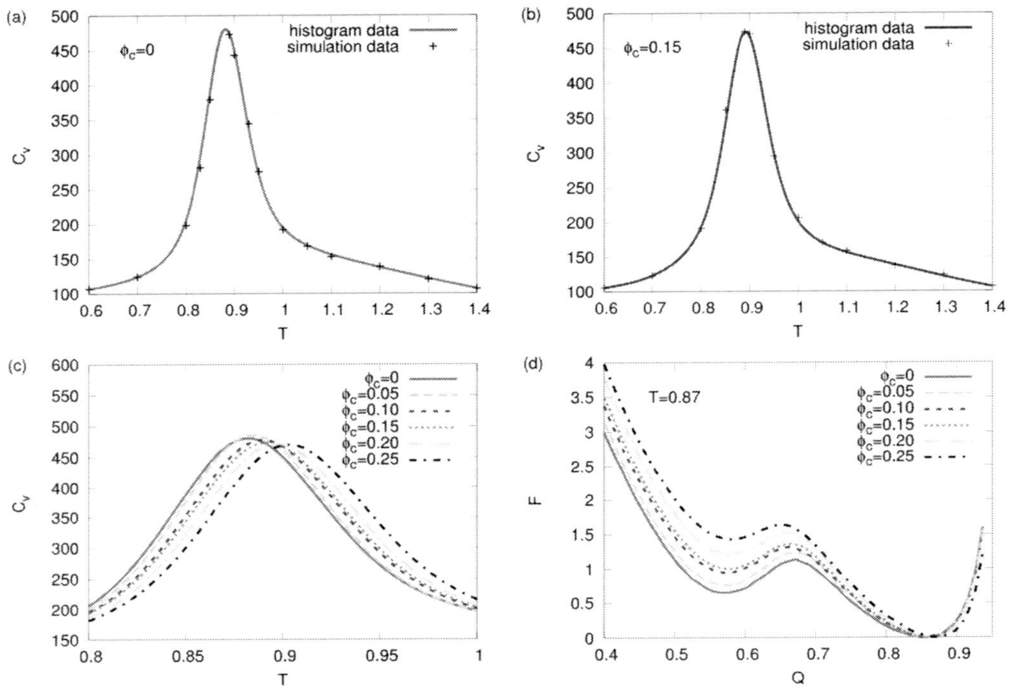

Figure 3. (a-b) Temperature dependence of the specific heat, C_v, of protein G in absence (a) and presence (b) of crowders plotted using non-weighted data from MD trajectories (crosses) and those obtained by the weighted histogram method (solid lines). (c) Comparison of the specific heats for different crowders' volume fraction $\phi_c = 0$, 0.05, 0.1, 0.15, 0.2 and 0.25 as indicated. (d) Dependence of effective free energy, F, on the fraction of native contacts, Q, at temperature $T = 0.87\epsilon/k_B$ for different values of ϕ_c as indicated. The data were obtained with crowder's radius $R_c = 10.2$ Å and container radius $R_{\rm wall} = 106.4\mathring{A}$.

Fig. 3 shows that for both the absence and presence of crowders the folding transition of GB1 is cooperative as characterized by a prominent peak in the specific heat and is clearly two-state as displayed by the two minima separated by a barrier in the free energy profile (Fig. 3d). The position of the specific heat's peak (corresponding to $T = T_f$) however is shifted towards higher temperature in presence of the crowders. As the crowders' volume fraction increases the folding temperature T_f also increases but the height of the specific heat peak somewhat decreases (Fig. 3c). These observations indicate that the crowders enhance the thermal stability of protein, but at the same time decreases folding cooperativity which usually is signified by the height of the specific heat peak. The crowders also enhance the thermodynamic stability of the native state or folding stability as can be seen from the free energy profile energy profiles (Fig. 3d)

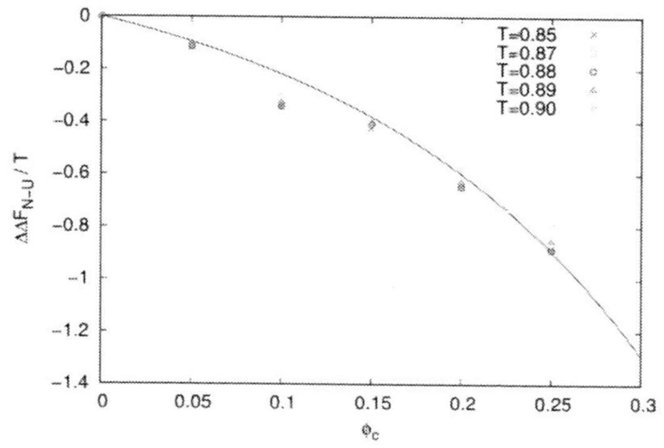

Figure 4. Dependence of the change of the folding free energy on crowders' volume fraction. The points shown are obtained by simulations for different temperatures as indicated whereas the solid line presents an approximate fit by scaled particle theory (SPT) with $R_c = 10.2\mathring{A}$, $a_N = R_g^N = 10.2\mathring{A}$, and $a_U = 1.13R_g^N$.

at a temperature below but close to the folding temperature. As ϕ_c increases the free energy difference between the unfolded state and the folded state increases. The free energy barrier for folding decreases which indicates that folding is faster in presence of the crowders. One can also notice that the position of the minimum corresponding to the unfolded state is shifted towards higher Q.

Scaled particle theory (SPT) predicts that the free energy of inserting a hard sphere of radius R in a fluid of hard spheres of radius R_c is given by [9]:

$$\beta F = (3y + 3y^2 + y^3)\rho + (9y^2/2 + 3y^3)\rho^2 + 3y^3\rho^3 - \ln(1 - \phi_c), \tag{10}$$

where $y = R/R_c$, $\rho = \phi_c/(1 - \phi_c)$ and ϕ_c is volume fraction of the fluid spheres. SPT has been used to estimate the effect of macromolecular crowding on the folding free energy of proteins, e.g. in Minton's theory [10], in which the folded and unfolded states are considered as effective spheres of radii a_N and a_U, respectively.

From the simulations, the folding free energy can be estimated as

$$\Delta F_{N-U} = F_N - F_U, \tag{11}$$

where F_N and F_U are the two free energy minima corresponding to the folded and unfolded states as found in the $F(Q)$ plot. The change in folding free energy due to the effect of crowders can be calculated as:

$$\Delta\Delta F_{N-U} = \Delta F_{N-U}(\phi_c) - \Delta F_{N-U}(0). \tag{12}$$

Fig. 4 shows that $\Delta\Delta F_{N-U}$ decreases monotonically with the volume fraction of crowders for a range of temperatures near the folding transition. The simulation data can be fitted quite well with SPT by using $R_c = 10.2\mathring{A}$, $a_N = R_g^N = 10.2\mathring{A}$, $a_U = 1.13R_g^N$.

4. Conclusion

In this study, we have shown that macromolecular crowding leads to an increased folding temperature and increased folding stability as defined by the free energy difference between the folded and unfolded states. The change in folding stability depends monotonically with the volume fraction of the crowders in good agreement with scaled particle theory. The enhancement of folding stability found here are consistent with experiments and current understanding on the effects of crowding on protein folding.

Acknowledgments
This research is funded by Vietnam National Foundation for Science and Technology Development (NAFOSTED) under grant number "103.01-2013.16".

References

[1] Minton AP 1983 *Mol Cell Biochem.* **55** 119
[2] Minton AP 2000 *Curr. Opin. Struct. Bio.* **10** 34
[3] Ellis RJ 2001 *Curr. Opin. Struct. Bio.* **11** 114
[4] Zhou HX, Rivas G, Minton AP 2008 *Ann. Rev. Biophys.* **37** 375
[5] Zimmerman SB 1993 *Biochim. Biophys. Acta.* **1216** 175
[6] Rivas G, Fernandez JA, Minton AP 1999 *Biochemistry* **38** 9379
[7] Van den Berg B, Ellis RJ, Dobson CM 1999 *EMBO J.* **18** 6927
[8] Van den Berg B, Wain R, Dobson CM, Ellis RJ 2000 *EMBO J.* **19** 3870
[9] Lebowitz J L and Rowlinson J S 1964 *J. Chem. Phys.* **41** 133
[10] Minton AP 2005 *Biophys. J.* **88** 971
[11] Zhou HX 2008 *Arch. Biochem. Biophys.* **469** 76
[12] Cheung MS, Klimov D and Thirumalai D 2005 *PNAS* **102** 4753
[13] Mittal J, Best RB 2010 *Biophys J.* **98** 315
[14] Cheung MS and Thirumalai D 2007 *J. Phys. Chem. B* **111** 8250
[15] Go N 1983 *Annu. Rev. Biophys. Bioeng.* **12** 183
[16] Hoang TX and Cieplak M 2000 *J. Chem. Phys.* **113** 8319
[17] Ferrenberg A M and Swendsen R H 1989 *Phys. Rev. Lett.* **63** 1195

IWTCP-2 and NCTP-39 IOP Publishing
Journal of Physics: Conference Series **627** (2015) 012028 doi:10.1088/1742-6596/627/1/012028

Aggregation of peptides in the tube model with correlated sidechain orientations

Nguyen Ba Hung[1,2] **and Trinh Xuan Hoang**[1]

[1] Institute of Physics, Vietnam Academy of Science and Technology, 10 Dao Tan, Ba Dinh, Ha Noi, Viet Nam

[2] Vietnam Military Medical University, 160 Phung Hung, Ha Dong, Ha Noi, Viet Nam

E-mail: hungnb80@yahoo.com.vn

Abstract. The ability of proteins and peptides to aggregate and form toxic amyloid fibrils is associated with a range of diseases including BSE (or mad cow), Alzheimer's and Parkinson's Diseases. In this study, we investigate the the role of amino acid sequence in the aggregation propensity by using a modified tube model with a new procedure for hydrophobic interaction. In this model, the amino acid sidechains are not considered explicitly, but their orientations are taken into account in the formation of hydrophobic contact. Extensive Monte Carlo simulations for systems of short peptides are carried out with the use of parallel tempering technique. Our results show that the propensity to form and the structures of the aggregates strongly depend on the amino acid sequence and the number of peptides. Some sequences may not aggregate at all at a presumable physiological temperature while other can easily form fibril-like β-sheet struture. Our study provides an insight into the principles of how the formation of amyloid can be governed by amino acid sequence.

1. Introduction

Amyloid fibrils are stable aggregates of proteins implicated in a range of diseases including Alzheimer's, Parkinson's and spongiform encephalopathies [1]. These fibrils have been found for a variety of proteins and synthetic peptides including those that do not involve in any disease [2, 3]. From these experimental evidences and simulations [4, 5, 6, 8, 9] it is generally accepted that that the tendency to form amyloid fibrils is a generic property of polypeptide chain. Yet, it is also believed that the amino acid sequence is a dominant factor in determining the propensity to form fibrils (see e.g. Ref. [10]) eventhough the role of sequence is much less known. Studies of the tube model of protein have shown that even homopolymeric peptides can aggregate and form fibril-like structures [5, 8, 9]. A general framework provided by tube model [4, 5, 7] also suggests that amyloid-like aggregates belong to a few minima in a presculpted free energy landscape of protein that is governed by symmetry of a tube-like polymer and geometrical constraints of hydrogen bonds. The role of amino acid sequence is to select or to avoid these minima.

In this study we investigate the role of sequence in the formation of amyloid by taking the advantages of the tube model in capturing the key features of polypeptide chain and of being strongly coarse-grained so that a full equilibrium characteristics of a pool of peptides could be studied. Furthermore, in the present study we incorporate sidechain orientations to the tube model in order to define better hydrophobic contacts. We will consider peptides of

Journal of Physics: Conference Series **627** (2015) 012028 doi:10.1088/1742-6596/627/1/012028

different hydrophobic-polar sequences but with the same hydrophobic fraction for making useful comparison.

2. Models and methods

We consider the tube model of protein [4, 11], in which each amino acid is considered as a single residue whose position is specified by the center of its C_α atom. The distance between two consecutive residues along the chain is 3.8 $\AA$. The residues are placed along the axis of a self-avoiding tube of cross-sectional radius $\Delta = 2.5\AA$. The finite thickness of the tube was imposed by requiring the radius of circle drawn through any three C_α atoms must be larger than Δ [12, 11]. Also, sterics requires that two non-consecutive C_α's cannot be closer than 4 $\AA$ from each other. The bond angles are restricted to the range of $82^o - 148^o$, and a bending stiffness is introduced by applying a local bending penalty energy of $e_R = 0.3\epsilon > 0$ if the local radius of curvature at a given bead is less than 3.2 $\AA$, where ϵ is an energy unit equal to the energy of a local hydrogen bond. Hydrogen bonds between amino acids are required to satisfy a set of distance and angular constraints [4] on the C_α's as found by a statistical analysis of PDB's native protein structures [5]. A local hydrogen bond is formed between residues that are separated by three peptide bonds along the chain, and is given an energy $-\epsilon$. A non-local hydrogen bond is given an energy of -0.7ϵ. To avoid spurious effects of the chain termini, hydrogen bonds (local and non-local) involving a terminal residue are given a reduced energy of -0.5ϵ. Additionally, a cooperative energy of -0.3ϵ is given for each pair of hydrogen bonds that are formed by pairs of consecutive amino acids in the sequence.

Hydrophobic interactions are based on pairwise contacts between amino acids. We assume that there are only two kinds of amino acids: hydrophobic (H) and polar (P). A contact between two non-consecutive beads is formed if they are found within a distance of 7.5 $\AA$. Only contacts between hydrophobic residues are favorable and are assigned an energy of $e_{HH} = -0.5\epsilon$ per contact. Contacts involving polar residues are given zero energy.

In the present study, we consider a modified tube model with a new procedure for hydrophobic interaction. The sidechains are not considered explicitly, but their orientations as given by anti-normal vectors are taken into account in hydrophobic contacts (Fig. 1). Apart from the distance constraint of 7.5 $\AA$ as described above, our new procedure requires that two residues i and j make a hydrophobic contact if $\mathbf{n}_i \cdot \mathbf{c}_{ij} < 0.5$ and $-\mathbf{n}_i \cdot \mathbf{c}_{ij} < 0.5$ where $\mathbf{n}_i$ is the normal vector associated with bead i, and $\mathbf{c}_{ij}$ is an unit vector pointing from bead i to bead j. The normal and connecting vectors are given by:

$$\mathbf{n}_i = \frac{\mathbf{r}_{i-1} + \mathbf{r}_{i+1} - 2\mathbf{r}_i}{|\mathbf{r}_{i-1} + \mathbf{r}_{i+1} - 2\mathbf{r}_i|} , \tag{1}$$

and

$$\mathbf{c}_{ij} = \frac{\mathbf{r}_j - \mathbf{r}_i}{|\mathbf{r}_j - \mathbf{r}_i|} , \tag{2}$$

respectively, where $\mathbf{r}_i$ is the position vector of bead i.

Monte Carlo (MC) simulations are carried out with periodic boundary condition. A parallel tempering [13] Monte Carlo scheme is employed for obtaining the ground state as well as other equilibrium characteristics of the system. For each system, 16 to 24 replicas are considered, each evolving at its own selected temperature T_i. For each replica, the simulation is carried out with standard pivot and crankshaft move sets and the Metropolis algorithm for move acceptance. An attempt to exchange replicas is made every 100 MC steps. The exchange of replicas i and j is accepted with a probability equal to $p_{ij} = \min\{1, \exp[k_B^{-1}(T_i^{-1} - T_j^{-1})(E_i - E_j)]\}$, where k_B is the Boltzmann constant, and E_i and E_j are the energies of the replicas at the time of the exchange.

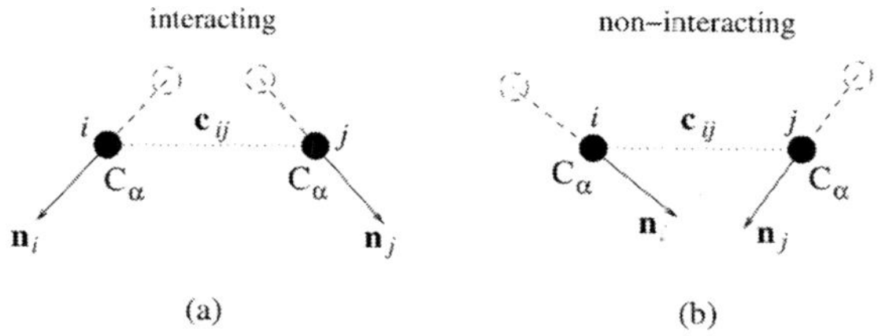

Figure 1. Contact interaction with sidechain orientation. The sidechain is approximately placed in the anti-normal direction from the C_α atom. Two residues are in contact if their sidechains are oriented not too apart from the other residue.

Table 1. HP sequences of amino acids of peptides considered in present study (H – hydrophobic, P – polar). The parameter s denotes sequence separation between two H amino acids.

Sequence name	Sequence	s
S1	P P P H H P P P	1
S2	P P H P H P P P	2
S3	P P H P P H P P	3
S4	P H P P P H P P	4
S5	P H P P P P H P	5
S6	H P P P P P H P	6
S7	H P P P P P P H	7

The temperature range in parallel tempering simulations are chosen such that it covers the transition from a dilute gas of separated peptides at high temperature to an aggregation phase at low temperature. The parallel tempering temperatures T_i are chosen such that acceptance rate of replica exchanges is significant, especially near the peak of the specific heat where energy fluctuation is large. Typically we need to change the set of temperatures several times in such a way that there are more temperatures for the replicas near the specific heat's peak. In each simulation, the number of Monte Carlo attempted moves for each temperature is of the order of 10^9.

The weighted multiple-histogram technique [14] is used to compute the specific heat of the system, which is given by:

$$C = \frac{\langle E^2 \rangle - \langle E \rangle^2}{k_B T^2} , \tag{3}$$

where E is the energy of the system and $\langle \cdot \rangle$ denotes thermodynamic average.

3. Results and discussion

We consider systems of $M = 10$ identical peptides of length $N = 8$ residues. The amino acid sequences of the peptides denoted as S1–S7 are given in Table 1. We have selected the sequences in a way that they have the same fraction of 25% of hydrophobic amino acids, i.e. each sequence has two H residues. The sequences differ in the separation between the two H residues which is given by the parameter s (see Table 1).

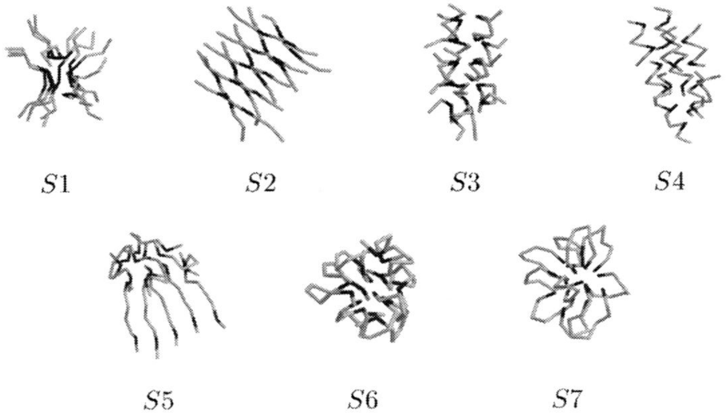

Figure 2. RASMOL visualization of aggregate structures for systems of 10 chains of the sequences (S1-S7) given in Table 1. H and P amino acids are shown in green and blue, respectively. The structures shown are lowest energy conformations obtained in parallel tempering Monte Carlo simulations.

The systems are simulated in a cubic box of box size $L = 255.15\mathring{A}$ with periodic boundary conditions. The box size was chosen such that the peptide concentration is equal to 1 mM. Figure 2 shows the lowest energy conformations obtained from parallel tempering simulations which are supposed to be the ground states of the systems. It is shown that the forms of the aggregates are remarkably different for different sequences. Only the sequence S2 system forms a double layer β-sheet structure which is similar to cross-β structures found in amyloid. α-helix bundles are seen for the S3 and S4 systems. Other sequences have disordered aggregates with the β-sheet appeared more clearly for sequence S5. Our results shows that the propensity to form amyloid-like structure strongly depends on the sequence.

Remarkable differences between sequences are also seen in the temperature dependence of the specific heats for the systems considered (Figure 3). Each sequence has its own specific heat profile with the temperature of the main peak varies between 0.17 and 0.26 ϵ/k_B and the heights of the peaks differ up to about one order of magnitude (Figure 4). The main peak in the specific heat signifies an aggregation transition and its height is a measure of the cooperativity of the transition. A sharp peak means that the transition is first-order like and the aggregate is an ordered structure. Note that the highest peak in the specific heat belongs to sequence S2 which forms the double layer β-sheet structure. This result suggests that the propensity to form amyloid may be linked to the cooperativity of the aggregation transition. The least cooperative aggregation transition belongs to sequence S1 which has two main peaks in the specific heat. For this sequence aggregation happens at the temperature of the higher peak but this transition just corresponds to a strong hydrophobic collapse and the aggregated structure remains disordered if the temperature is higher than the temperature of the lower peak.

The differences in the transition temperatures T_{max} (corresponding to the maximum of the specific heat) among sequences suggest a very interesting consequence. Suppose that we are interested in the behavior of systems at the physiological temperature, T^*. A rough estimate of T^* in our model is about 0.2 ϵ/k_B (which corresponds to an energy of a local hydrogen bond of 5 k_BT with T is room temperature). With this estimate from Figure 4B one can see that out of 7, three systems corresponding to sequences S3, S4 and S5 have the transition temperature T_{max} lower than the physiological temperature T^*. Thus at $T = T^*$ these three systems do not

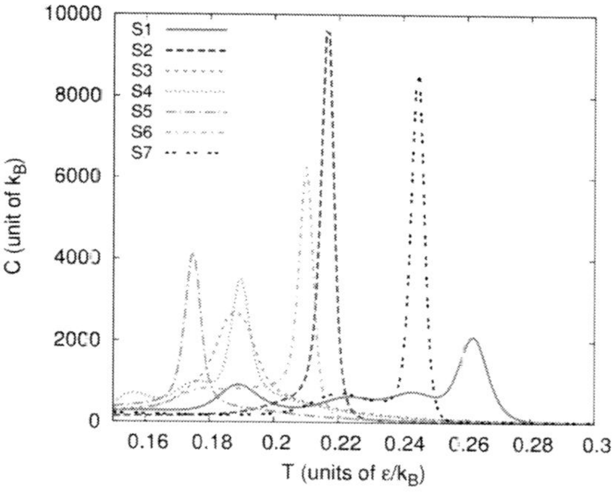

Figure 3. Temperature dependence of the specific heat C for systems of 10 peptides of HP sequences as given in Table 1 (S1–S7).

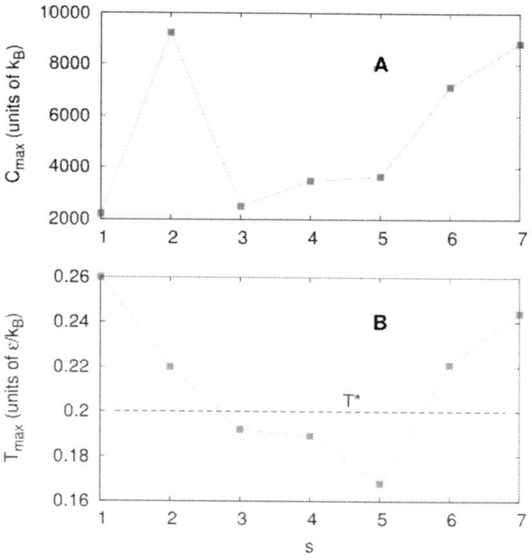

Figure 4. Dependence of the maximum of the specific heat C_{max} (A) and its temperature T_{max} (B) on the sequence separation s between two H residues in the HP sequence for the 7 systems considered. The horizontal dashed line in (B) indicate a presumed physiological temperature T^*.

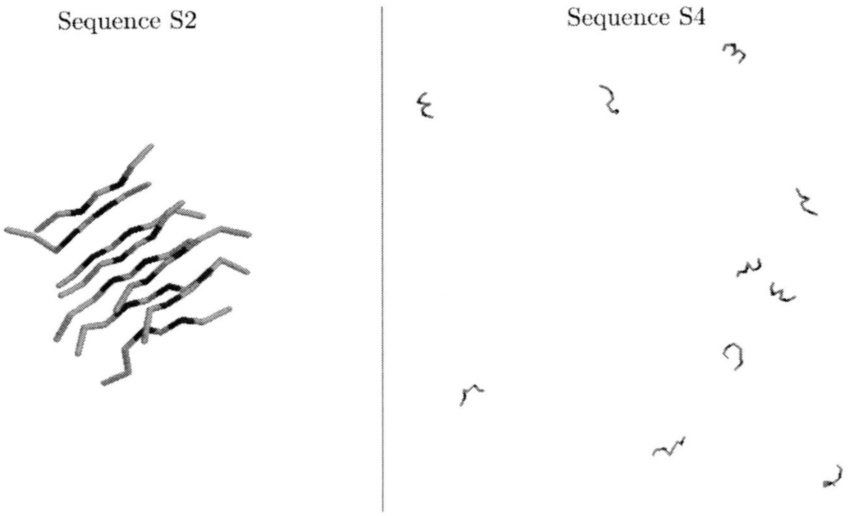

Figure 5. Snapshots of conformations at $T = 0.2\epsilon/k_B$ for peptides of sequence S2 (left) and sequence S4 (right). Note that at the given temperature the S2 peptides aggregate and form an amyloid-like structure whereas the S4 ones do not aggregate.

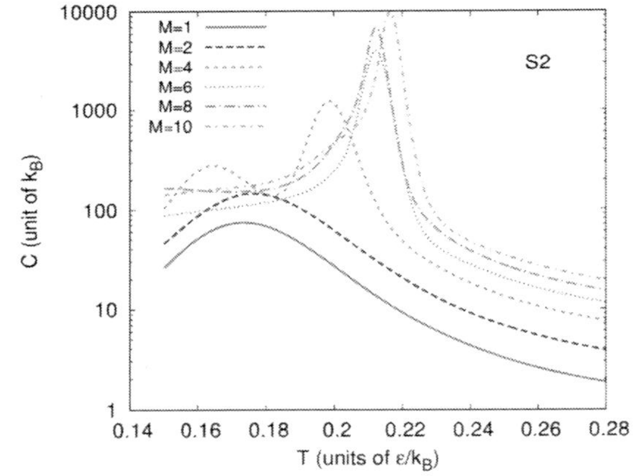

Figure 6. Temperature dependence of the specific heat for sequence S2 with various numbers of chains, M, as indicated.

aggregate while the other systems including that of sequence S2 do. This situation is illustrated in Figure 5 which shows snapshots obtained at $T = 0.2\epsilon/k_B$ for S2 and S4 systems. One can see that the S2 system has already formed a amyloid-like two-layer β-sheet structure while the S4 system has not aggregated. About 50% of chains in the S4 system form stand-alone α-helices.

Figure 6 shows the temperature dependence of the specific heat for increasing number of chains for sequence S2. It is shown that the height and the position of the peak increases as M increases which suggests that the propensity of peptides to aggregate grows with the density of peptides. For single chain the ground state of this sequence is an α-helix. The amyloid-like β-sheet structure starts to form when $M > 4$. This result agrees with the nucleation and growth mechanism of amyloid formation [3, 8, 9]. For the case of $M = 4$, the specific heat has two peaks that correspond to a collapse transition at the higher temperature and a disorder-oder transition at the lower temperature.

In a recent study [15] Dobson and coworkers considered a lattice model with side chain directions for amino acids to study amyloid formation of short peptides. Their designed sequence that was shown to form fibrillar structure is remarkably similar to sequence S2 in our model. Our present study confirms the important role of sequence in amyloid formation in a more realistic model. Two features of the sequence that are important for amyloid formation are the parallel pairing of hydrophobic sidechains within a cross β-sheet and the anti-parallel facing of hydrophobic sidechains between β-sheets in a fibirilar structure.

4. Conclusion

We have studied aggregation of peptides by performing Monte Carlo simulations using a modified tube model with correlated sidechain orientations. We have shown that the structure of the aggregates strongly depends on the HP sequence of peptides, and so does the aggregation transition as indicated by the specific heat. In particular, for a sequence with two H amino acid separated by one P, the aggregate is a two-layer β-sheet structure similar to that found in amyloid fibrils. The aggregation transition for this sequence is also the most cooperative and first-oder like among all sequences. Interestingly, our simulations indicate that some sequences have a very low propensity to form amyloid and they may not aggregate at all at a presumable physiological temperature. We conclude that the hydrophobic-polar patterns in the amino acid sequence play an important role in protein aggregation and they could be a dominant factor in determining the propensity of a polypeptide chain to form amyloid fibrils.

Acknowledgments

This research is funded by Vietnam National Foundation for Science and Technology Development (NAFOSTED) under grant number "103.01-2013.16".

References

[1] Dobson C M 1999 *TIBS* **24** 329
[2] Chiti F, Dobson C M 2006 *Annu Rev Biochem.* **75** 333
[3] Hellstrand E, Boland B, Walsh D M and Linse S 2010 *ACS Chem. Neurosci.* **1** 13
[4] Hoang T X, Trovato A, Seno F, Banavar J R and Maritan A 2004 *Proc. Natl. Acad. Sci. USA.* **101** 7960
[5] Banavar J R, Hoang T X, Maritan A, Seno F and Trovato A 2004 *Phys. Rev.* E **70** 041905
[6] Hoang T X, Marsella L, Trovato A, Seno F, Banavar J R and Maritan A 2006 *Proc. Natl. Acad. Sci. USA.* **103** 6883
[7] Banavar J R and Maritan A 2007 *Annu. Rev. Biophys. Biomol. Struct.* **36** 261
[8] Auer S, Dobson C M and Vendruscolo M 2007 *HFSP Journal.* **1** 137
[9] Auer S, Dobson C M, Vendruscolo M and Maritan A 2008 *Phys. Rev. Lett.* **101** 258101
[10] Li M S, Co N T, Reddy G, Hu C K, Straub J E and Thirumalai D 2010 *Phys. Rev. Lett.* **105** 218101
[11] Maritan A, Micheletti C, Trovato A and Banavar J R 2000 *Nature.* **406** 287
[12] Gonzalez O, Maddocks J H 1999 *Proc. Natl. Acad. Sci. USA.* **96** 4769
[13] Swendsen R H, Wang J S 1986 *Phys. Rev. Lett.* **57** 2607
[14] Ferrenberg A M, Swendsen R H 1989 *Phys. Rev. Lett.* **63** 1195
[15] Abeln S, Vendruscolo M, Dobson C M and Frenkel D 2014 *PLOS One* **9** e85185

Institute of Physics
Dirac House, Temple Back
Bristol BS1 6BE UK

CURRAN ASSOCIATES INC
proceedings.com